SECRET JOURNEY TO PLANET SERPO

A True Story of Interplanetary Travel

Len Kasten

Bear & Company
Rochester, Vermont • Toronto, Canada

Bear & Company
One Park Street
Rochester, Vermont 05767
www.BearandCompanyBooks.com

Text stock is SFI certified

Bear & Company is a division of Inner Traditions International

Library of Congress Cataloging-in-Publication Data
Kasten, Len.
Secret journey to planet Serpo : a true story of interplanetary travel / Len Kasten.
p. cm.
Summary: "Documents how 12 people, as part of a top-secret U.S. government program, traveled to the planet Serpo and lived there for 13 years" — Provided by publisher.
ISBN 978-1-59143-146-6 (pbk.) — ISBN 978-1-59143-831-1 (e-book)
1. Extraterrestrial beings. 2. Conspiracies—United States. 3. Interplanetary voyages. 4. Human-alien encounters. I. Title.
TL789.K3565 2013
001.9—dc23

2012038164

Printed and bound in the United States by Lake Book Manufacturing, Inc.
The text stock is SFI certified. The Sustainable Forestry Initiative® program promotes sustainable forest management.

10 9 8 7

Text design by Jon Desautels and layout by Virginia L. Scott Bowman
This book was typeset in Garamond Premier Pro and Avenire with Stencil used as the display typeface

To send correspondence to the author of this book, mail a first-class letter to the author c/o Inner Traditions • Bear & Company, One Park Street, Rochester, VT 05767, and we will forward the communication, or contact the author directly at **www.et-secrethistory.com/index.html**.

CONTENTS

PART THREE
EPILOGUE

APPENDICES

Sparks (little girl): Hey Dad, do you think there's people on other planets?

Dad: I don't know Sparks, but I guess I'd say if it is just us, seems like an awful waste of space.

Contact (Warner Bros. Films, 1997)
Screenplay by John V. Hart and Michael Goldenberg
Based on the book *Contact* by Carl Sagan

INTRODUCTION

Queen Elizabeth I: *A play! Comedy or tragedy?*
Player: *Comedy, your Majesty.*
Queen Elizabeth I: *Comedy! By whom?*
Player: *Anonymous, your Majesty.*
Queen Elizabeth I: *Anonymous! I so admire his verse.*

ANONYMOUS,
COLUMBIA PICTURES (2011);
SCREENWRITER: JOHN ORLOFF

ANONYMOUS AND PROJECT CRYSTAL KNIGHT

> First let me introduce myself. My name is . . . Anonymous. I am a retired employee of the U.S. government. I won't go into any great details about my past, but I was involved in a special program.

With those words, sent in an e-mail to Victor Martinez, the host and moderator of perhaps the largest and most prestigious space-related e-mail network on the Internet, known as the UFO Thread List, a totally new era of government transparency commenced. It was sent to Martinez on November 2, 2005, followed by incredible facts about what the government knows about alien visitations. "Anonymous" then sent in eighteen more e-mails to the network, each more sensational than the last. That first batch continued through August 21, 2006. He then sent in fourteen more e-mails between June 4, 2007, and April 13, 2011, revealing heretofore top-secret information that had previously been classified as "Top Secret Codeword," the highest secrecy classification in the government.

The e-mail postings by Anonymous mainly concerned the events following the Roswell crash (see chapter 3) that led to an interstellar exchange program in which we sent twelve U.S. military personnel to a distant planet on an alien spaceship in 1965. This program was directed and monitored by the Defense Intelligence Agency (DIA). The DIA called the program Project Crystal Knight, and Anonymous, then much younger, was allegedly one of the agency officials assigned to the project. It is now popularly known as Project Serpo.

CREDIBILITY, "THE LIST," AND WWW.SERPO.ORG

That first e-mail to the network immediately elicited comments by long-time network members. Gene Loscowski (real name: Gene Lakes) said, "Who is this person? Most of the information is absolutely correct." Paul McGovern said, "Interesting, but not totally correct." List member "Anonymous II" said, "As for the Roswell incident: This was the story I read in the historical document called the 'Red Book.' Almost exactly to the word. Although there were more details about the crash sites and what was recovered."

These remarks from individuals who apparently knew at least some of the details of the secret program, should be considered definitive confirmation that Project Crystal Knight was a real event and that the story that Anonymous revealed was substantially correct. It should be noted that virtually all the people on the list were considered "insiders" in one way or another and not just a collection of people with an interest in UFO activities, whether they be investigators or even abductees. These individuals were largely government connected and many had been exposed to classified information. That is precisely why Anonymous chose to make these momentous revelations to this group. Bill Ryan, who later published the website for the list, explained it this way:

> The list at that time contained about a hundred and fifty people, including many extremely well-known names in UFO research and related or leading-edge scientific fields. . . . Those on the list have differing views regarding the veracity of Anonymous's claims. However, the pedigree of the list as a whole is important to emphasize. There has been a substantial amount of intelligent discussion about the revelations, and *it is important to state that there are many senior people in the U.S. intelligence and military community who are taking this information very seriously.*

Any skepticism about the Serpo revelations by Anonymous among list members largely evaporated as he sent in more and more details about the program to the list network. It soon became very clear that this amount of detail, some of which they could personally attest to, or had already heard about, could not have been invented. Consequently, on December 21, 2005, network members decided to launch a website dedicated to the revelations of Anonymous. To that end, a longtime, well-respected British member of the network named Bill Ryan volunteered to create and moderate the website, named www.serpo.org. That website ultimately became fleshed out with additional material from other anonymous contributors and from postings by Martinez, who added supplementary explanations relating to the information revealed by Anonymous and other insiders. The result was an amazing compendium of government knowledge about our relations with aliens from across the galaxy.

THE *RED BOOK*

Any lingering doubts about the credentials of Anonymous and his role in Project Serpo were completely erased by a post that he sent in to the website on June 16, 2006. In that e-mail he identified himself as the editor of the *Red Book.* Since he spoke in the present tense, we must

Serpo website developer Bill Ryan

conclude that he still acted in that capacity in 2006. This mysterious "book" is well-known to highly placed government officials involved with secret UFO investigation and contact with extraterrestrials. Anonymous describes the *Red Book:*

> The Red Book is an extremely thick, very detailed account summary, written and compiled by the U.S. government on UFO investigations dating from 1947 to the present day. This orange-brownish book is updated every five years.

Then, on August 9, 2007, Anonymous sent the following message, elaborating on the use of the *Red Book* and revealing his role in producing and editing the entries in the book:

> . . . claims about the Red Book being updated continuously or when needed are somewhat true, but need to be placed in proper context.
>
> What actually occurs is that as UFO reports come in [and are] deemed credible by the reporting government agency—be it military or civilian—they are routed to a special section of our government for a follow-up analysis. After that vetting process, they are then sent to a special group which then places [them] into final review for POSSIBLE inclusion into the Red Book.

. . . I know all of this because . . . I have served as the editor for several editions of the Red Book and have written and delivered the Executive Summary for several sitting U.S. presidents, so I KNOW of what I speak. And when I say "editor," it is NOT in the sense of the word you are familiar with. I do not correct nor review any of the hundreds—if not thousands—of reports which are distilled into the final five-year report for grammar and punctuation as you've done with the "Project Serpo" material.

I only present and include the most important, compelling cases [in] the RED BOOK, which includes an analysis by me and others of any trends, types of sightings, human contacts with the ETEs [extraterrestrial entities] and any national security concerns our government or planet might have. My part is to write the Executive Summary and present it to the current sitting president of the United States. If there was a national security matter that presented itself, that five-year published review of the RED BOOK would be interrupted, but that has been unnecessary as we have a good relationship with our visitors [aliens].

We did have visitors from nine other star systems. The Grays, which some people characterize as being like the Ebens, were not [Ebens]. They came from a planet near Alpha Centauri A. The third class of visitors came from a G2 star system in Leo. Another class of visitors came from a G2 star system in Epsilon Eridani. The visitors were classified by a code. The code, which was classified in itself, was, "Extraterrestrial Entities" (ETE). ETE-2 were the Ebens, the Grays were ETE-3, and so on. The "Red Book" lists nine different visitors. We determined recently that some of the visitors were the same type of race but a "mechanical life-form." They were hybrid beings that were created in a laboratory rather than by natural birth. The creatures were more like robots, although they were intelligent and could make decisions. They might be the "hostile" visitors that some people report.

Anonymous's involvement with the *Red Book* and with briefings to the presidents clearly places him at the upper reaches of the intelligence community with regard to relations with extraterrestrials. Although he never mentions it, it seems very likely that he is, or has been, a member of Majestic 12 (MJ-12), the secret organization set up by President

Harry S. Truman to deal with extraterrestrial affairs. Consequently, his Serpo revelations should be considered authentic.

ANONYMOUS AND THE DIA

In further communications to the network, Anonymous discloses that he is not acting entirely alone, but is part of a group of individuals from the DIA. In his Introduction to the website, Bill Ryan says:

> Anonymous reports that he is not acting individually and is part of a group of six DIA personnel working together as an alliance: three current and three former employees. He is their chief spokesman. . . . When sending information to Victor Martinez, Anonymous wrote 85 percent of the material; another 13 percent came from another source directly connected with the project; and the final 1–2 percent came from a "ghost," who canceled his e-mail account as soon as he sent his information.

AUTHORIZATION

The revelations made by Anonymous and the DIA-6 were not some sort of rogue operation, but were authorized for release at the highest level within the DIA. It's not known whether or not that authorization came from an even higher level within the U.S. intelligence community and the executive branch, or from MJ-12. As is pointed out in chapter 6, the DIA was created by President John F. Kennedy in part to make the intelligence network more accountable to the public, rather than permitting it to maintain a policy of exclusivity and frequently high-handed behavior. Consequently, the ideal of transparency was instilled in the DIA from the beginning, and this ideal has remained in its DNA. These revelations by the DIA-6 reflect that ideal. As quoted by Victor Martinez in his brief history of the DIA (see appendix 8), a Pentagon official told *The Washington Post*: "We've got to do to intelligence what CNN has done to news."

The reader can accept Anonymous's writings as genuine with confidence, because it seems highly improbable that he, now at an advanced age and retired, should suddenly decide to advance a piece of very complicated government propaganda as part of a disinformation campaign. What would be the point of creating and telling this fantastic story, almost in the realm of science fiction, at this point in history and in his life, when he didn't need to make those disclosures? The fact that all this information was written in the *Red Book* would strongly tend to nullify the disinformation hypothesis. The enormous amount of detail alone makes the likelihood that this whole story was the product of his imagination almost impossible. It would place him among the science-fiction ranks of Jules Verne, H. G. Wells, and Isaac Asimov!

WHY NOW?

The hypothesis that is far more likely is this: an aging, highly placed intelligence operative in an agency inspired by the transparency policy instilled by President Kennedy has decided that, as he nears the end of his life, the public has a right to know the fantastic truths about our dealings with extraterrestrials. It is a matter of "Damn the torpedoes, full speed ahead," and "Tell the truth, and let the chips fall where they may." Anonymous chose the date of November 2, 2005, because it is precisely twenty-five years after the final report on Project Crystal Knight was written (1980). Government policy allows for the declassification and disclosure of secret documents after that quarter-century period. One can only remark, along with Queen Elizabeth I, "I so admire his verse."

THE BACKGROUND

In telling the Serpo story, it seemed to me that it was necessary to first talk in part 1 about events that preceded the 1947 Roswell crash so that the reader is not left with the impression that Roswell was our first exposure to antigravity spacecraft or to extraterrestrials. In fact, the U.S. military had been dealing with these types of craft for at least five years at the time of Roswell. Furthermore, and most importantly, we had been aware of the alien presence on the planet since the 1930s, and we had learned about their role in World War II.* So, to the military at least, Roswell did not really engender any form of culture shock. It seemed to me that it was important to give some of this pre-Roswell history, not previously available to civilians, so that the reader can understand the state of mind of our military leaders in July 1947, and why they reacted with such alacrity to the crash and to the hosting of the surviving alien. In spite of appearances, they were not terribly surprised by the whole series of events, and, in fact, were probably better prepared for it than they were for Pearl Harbor on December 7, 1941. The Pentagon knew immediately what it meant to have an alien craft crash on U.S. soil. Military leaders recognized all the implications thereof, and what sort of culture shock could be expected in the American populace. So part 1 complements the Serpo saga, and together they paint a realistic picture of America's introduction to galactic affairs in the twentieth century. It is an amazing story, far more incredible than science fiction. And yet it is only the first chapter. What awaits the human race in the twenty-first century cannot even be imagined now, in our wildest dreams.

THE SERPO SAGA

In part 2 of the book, we learn of the amazing adventures of a band of twelve brave and resolute Americans who were willing to leave the comforts and familiarities of their home planet for a stellar destination in the far reaches of space that they knew almost nothing about! This level of bravery has no real precedent. It could perhaps be equated with the daring of the Spanish Conquistadors who landed in a new world—comparable to a new planet—and also faced an alien race. But those men were bolstered by the belief that they had nothing to fear from a race that they believed to be primitive savages. The Serpo twelve were

*According to Philadelphia Experiment survivor Al Bielek, President Franklin D. Roosevelt met with Pleiadian representatives onboard the USS *Missouri* in mid-Pacific in 1933. Bielek claimed that this meeting was arranged by Nikola Tesla.

willing to step off planet Earth and face unknown hardship and danger living in the midst of a civilization of aliens who looked, acted, and thought differently from Earthlings, and who were obviously more intelligent since they had the technology to travel across space. They would be living in a world where nothing was familiar and where they would be deprived of the consolations of their human friends and lovers, who were now across an ocean of space and with whom there could be no communication.

It would be a mistake to attribute their bravery simply to a sense of adventure, to say that their fear and trepidation was trumped by the excitement of experiencing living in such a new and startlingly different world. This kind of bravery had to be rooted in a determination to advance human knowledge about life in our galaxy. Somewhere in their minds there had to be those small voices that said that the time had come for the human race to push out into space, similar to that same voice in the mind of Christopher Columbus that told him that humans must explore this entire planet that we dwell upon. So these twelve were very special people. It is difficult to even imagine such courage and audacity. And yet, we don't even know their names! The government has rigidly maintained the policy that they must remain anonymous. Someday, somewhere, almost certainly, there will be a monument built to honor these intrepid space pioneers. Perhaps it will be at NASA, or in Houston, or at Cape Kennedy. But it really belongs at the Mall in Washington D.C., where it will remind tourists and visitors for all time that twelve courageous Americans dared to travel across the galaxy to a distant star system to advance the human race to a new level of knowledge and experience.

EBEN TECHNOLOGY AND *CLOSE ENCOUNTERS*

In the final section of the book are addressed two subjects that will be of major interest to all who have followed the story to this point. The first has to do with Eben technology. Since learning about their science and technology was probably the major reason for sending the Team to this distant planet, it is necessary that this subject be covered in detail and that it be compared to our earthly science and technology. Hence, it was believed necessary that we give this important subject its own chapter.

And finally, it is well known by now that Steven Spielberg's classic film, *Close Encounters of the Third Kind* contains a final scene showing the departure of twelve American military personnel on an alien spaceship to their home planet. Consequently, a reader of this book is likely to ask whether that scene was actually true and portrayed the Serpo team leaving Earth on an Eben craft. We considered that question important enough to devote an entire chapter to comparing the film to the real story. The conclusion is surprising.

THE APPENDICES

The appendices form a very important part of this story. This is detailed information about the journey, much of it too technical to be included in the main body of the book. It was necessary to supplement the information conveyed in the narrative with facts and figures and background to what could only be covered summarily in part 2, so as not to clog those chapters with an unwieldy number of footnotes. I have included thirteen appendices, all of which I consider indispensable to the full understanding of this remarkable epic adventure. It is in this section that we learn such details as the fact that the Team brought with them recordings of the music of The Beatles and Mozart, as well as three Jeeps and twenty-four handguns, among the forty-five tons of supplies and equipment. It is in this section where we present the entire verbatim briefing given to President Ronald Reagan about Project Serpo by the CIA. Appendix 10 can be read as a firsthand account of how, with the assistance of the Ebens, we have reverse-engineered alien craft beginning in 1953. Perhaps most important of all is the amazing "true confession" of the goals of the Public Acclimation Program in appendix 12. This document, simply titled "A Framework," confirmes what every UFO investigator and researcher since 1947 has believed, but which could never before be revealed. In effect, this brief summary constitutes the full disclosure for which Ufologists have been beleaguering and harassing the government for sixty-five years. It is really the "Holy Grail" of UFO and ET related research! Essentially, once the complete story of Project Serpo was released, these goals could really no longer remain hidden. And so, in an admirable and remarkable gesture of government transparency and honesty for the ages, here are the full admissions we have all been waiting for.

THE BOOK VERSUS THE WEBSITE

A website is an evanescent form of communication: it exists solely at the discretion of its creator, and can disappear literally overnight. If www.serpo.org were to be taken down—due to a change of transparency policy, a change of executive administration, or even on a whim—the world would be left without any record of this incredible event, and all the earnest work and dedication of these heroic DIA personnel would be lost to posterity. That is why it was so important to get this information into an indelible medium—that is, book form—as quickly as possible. That has now been accomplished in this volume.

PART ONE

PRELUDE

Before presenting the Serpo story, it is necessary to convey some preliminary information about important events that preceded the actual voyage. Knowing about these events will give the reader the historical background necessary to understand why MJ-12, the president, and the Pentagon were willing to send twelve American Air Force astronauts to a distant star system on an alien spaceship in 1965. While the events in Germany and Antarctica in the 1930s and 1940s may seem remote from American government postwar space-related decisions, there is a critical connection. After the war, the U.S. military was justifiably alarmed about the Nazi development of antigravity fighter discs housed in their secure ice-bound redoubt in Neuschwabenland in Antarctica. Thanks to the intelligence that Admiral Richard Byrd brought back from Operation Highjump in 1947, the Pentagon knew that the U.S. was vulnerable to an invasion by these craft, and that we were basically defenseless against such an assault. American fighter jets could not possibly be successful against aircraft that could hover, fly at supersonic speeds, and change direction instantly. And we knew from British intelligence that there were large submarines and a sizeable army billeted in Neuschwabenland, so the Germans could also mount an effective land offensive as a follow-up to a devastating air attack. Consequently, when the friendly Eben aliens from Zeta Reticuli offered a diplomatic exchange program that would expose us to information about their technology—far ahead of anything the Germans possessed—we simply could not refuse such an invitation. Because of the Nazi colony at Neuschwabenland and the space technology that we knew they were developing there, as well as the possibility that Hitler himself continued to command their forces, World War II was not yet over, and it was necessary for the American military to quickly adapt to interplanetary realities in order to win. Because Eben scientists were already here and helping us to develop antigravity technology, it made perfect sense for us to visit their planet and to learn whatever we could about this other remarkable civilization that shared our habitation in the Milky Way Galaxy.

1

GERMANY

While it is popularly believed that the Roswell incident in July 1947 (see chapter 3) was our first encounter with antigravity discs, or "flying saucers," that is not the case. Our military was very familiar with this phenomenon, having knowledge of and experience with these craft extending back to World War II. It is necessary to cover this experience in depth in order to fully comprehend the state of mind of the U.S. military at the time of the Roswell crash.

It is now well known that German aeronautical engineers were working on antigravity discs beginning in 1933, and by 1945 they had developed a highly sophisticated circular craft that was able to fly at very high altitudes and with incredible speed using electromagnetic propulsion technology. If the war had lasted only a few months longer, the Germans would have found a way to adapt these craft to aerial warfare, and the Allies' air superiority would have vanished. We would then have lost the war because our main advantage was control of the air. Even though this research and development was contained within the SS, and consequently was ultrasecret, thanks to deeply embedded Allied spies in the Nazi ranks, both Britain and the United States were receiving reports about these discs throughout the war. General Eisenhower was fully informed, as was Winston Churchill. Consequently, it became imperative to conclude the war as quickly as possible. It was very fortunate for the Allies that Hitler chose to fight on two fronts, which made an early victory possible. When he declared war on the United States on December 11, 1941, his armies in the East had not yet experienced a Russian winter, and he was still confident of a quick conquest of the Soviets. When that didn't happen, thanks largely to the American and British armaments pouring into Russia through Murmansk, and the heroic Soviet defense of Stalingrad, as well as the frigid temperatures, the previously invincible German troops were caught in a giant Allied pincer movement following D-Day, and defeat followed swiftly.

KARL HAUSHOFER

The central figure involved in obtaining the antigravity technology for the Nazis was a celebrated veteran of World War I who had emerged from that conflict with a lustrous reputation, despite the fact that the Germans lost the war. Karl Ernst Haushofer became a professional soldier at the age of eighteen in 1887 and completed artillery school and officer training at the War Academy of the Kingdom of Bavaria. In 1896, he married Martha Mayer-Doss, whose father was Jewish. He then rose through the ranks in the Imperial German Army until in 1903, at thirty-four, he became a teacher at the War Academy. During that period, the German-Prussian military enjoyed great foreign prestige as a result of their victory over the French in the Franco-German war in 1871. After teaching at the War Academy for five years, he was sent to Japan in 1908. Based in Tokyo, he was expected to study Japanese military practices and to act as an artillery instructor for the Japanese army. Japan's Imperial

Karl Haushofer

Army, which became centralized under the emperor at the outset of the Meiji era in 1871, was modeled after the Prussian army, and had already brought in French, Italian, and German advisors in the early stages.

According to uncredited information on Wikipedia:

> By the 1890s, the Imperial Japanese Army had grown to become the most modern army in Asia, well-trained, well-equipped with good morale. However, it was basically an infantry force deficient in cavalry and artillery when compared with its European contemporaries. Artillery pieces, which were purchased from America and a variety of European nations, presented two problems: they were scarce, and the relatively small number that were available were in several different calibers, causing problems with their ammunition supply.

Hence, Haushofer was brought to Japan in 1909 based on his artillery expertise, and his general experience with Prussian military discipline. He was greeted by the emperor upon his arrival, and he enjoyed a highly privileged social status while he and his family remained in Japan. Already fluent in Russian, French, and English, he easily added Japanese and Korean to his language repertoire, and, consequently, he was accepted into the highest strata of Japanese society and mingled with the power brokers in the emperor's circle. It was in this circle that Haushofer encountered the secret society at the base of Japanese political power, which encompassed the emperor as a figurehead—the Kokuryu-Kai, better known as the Black Dragon Society. Ultranationalistic, militaristic, and fascist, the Black Dragons, in addition to controlling Japan, infiltrated the power centers of all the countries of East Asia, even extending to the United States. They did not hesitate to use assassination and propaganda to further their goal of Japanese world hegemony.

DRAGONS—BLACK AND GREEN

At the innermost core of the Black Dragon Society was the Green Dragon Society. Here, political and economic power resolved down to occult and black magical power. Ostensibly, the Green Dragons were a small Buddhist monastic sect, although the monks also observed the Shinto ceremonies. In the sixteenth century, they chose Kyoto as their central location. In the nineteenth century it became known that the Green Dragons maintained a close affiliation with a mysterious group called the Society of Green Men, who lived in a remote monastery and underground community in Tibet and communicated with the Green Dragons only at the astral level. Capable of manifesting tremendous psychic and occult power, the Green Men easily controlled the Green Dragons, who viewed the liaison as advantageous to them, not realizing who was controlling whom.

Able to see into and travel through time, the Green Men had far-reaching plans extending to the year 5000. Impressed with German-Prussian militarism, the Green Men decided that an alliance with Germany would help them reach their fiftieth-century goals. Consequently, they convinced the Green Dragons to invite Haushofer into the society, and to initiate him into some of their mysteries. By giving him powers that only they could convey, they hoped to use him as the catalyst to bring about an invincible fascist German state,* which would form an alliance with Japan. Then, together, they would conquer Russia, and would then rule the massive Eurasian landmass and be in a position to confront the West European and Anglo-American alliances. Japan had already defeated Russia in the Russo-Japanese war of 1904–1906, which was largely a naval battle but gave them control over the highly strategic Port Arthur, which became a Japanese foothold in

*The philosophy of fascism is predicated on a homogeneous racial population binding all citizens into one organism so that the state becomes supreme and the individual amounts to little or nothing. This appears to be an attempt to duplicate, in human politics, the so-called hive mentality of the aliens (Greys), and is clear evidence of the role of the extraterrestrials in influencing the development of fascism in the Axis nations. Since individual initiative counts for nothing, the entire burden for leading a fascist state falls upon the dictator. For this reason, it is believed to be the ideal governmental philosophy for conducting war, since it eliminates the indecision that results from political squabbling. However, World War II exposed this as a fallacy, since it proved that democracies, relying on diversity, are much more resourceful and able to bring more varied initiatives and intelligence into the conduct of war.

Ryohei Uchida, founder of the Black Dragon Society

Russian-occupied Manchuria. So the Black Dragons were confident that they would be able to conquer Russia with Germany's help by attacking from two directions.

Haushofer became only the third Westerner ever to be initiated into the Green Dragon Society. He returned to Germany in 1911. Now forty-two years old, he was a changed man. Essentially, he was a secret weapon unleashed upon unsuspecting German society by the dugpas, the black magicians of the underground Tibetan civilization to further their visionary, global strategy for building a fiftieth-century world empire. He was, in effect, a missile aimed at the heart of European politics, although it is highly probable that he himself did not understand the true nature of his mission. More than likely, he had been hypnotized and brainwashed by the Green Dragon monks to believe in what he was doing. Upon his return to Germany, Haushofer had several illnesses and was unable to work for three years. However, he was well enough during that period to obtain his doctorate in geopolitics from Munich University. His doctoral thesis was titled "Reflections on Greater Japan's Military Strength, World Position, and Future," testifying to his continuing obsession with Japan. In 1914, he entered World War I with the rank of general, and was put in charge of a brigade on the Western front. His successes in the war became well known, as he was able to predict with precision Allied bombardments and maneuvers, and consequently, was able to undertake timely countermeasures, thus clearly demonstrating the powers of foreknowledge conferred upon him by the Green Dragons. As a result, he emerged from the war as a hero in the rank of major general and was highly respected in postwar German society.

SNAKEWORLD

The exact location of the Tibetan monastery of the Society of Green Men is not known, and it is unlikely that Haushofer knew where it was. The monks communicated with the Green Dragons through the astral realm, so they never had to reveal where they were. In retrospect, it has become clear that the Green Men were connected with the huge underground empire of the Reptilian extraterrestrials from Alpha Draconis, which is said to extend from southwestern Tibet all across the Indian subcontinent to Benares, India. This empire is called Patala, or "Snakeworld" in Hindu mythology, and is said to be the home of the storied Nagas, or the serpent race, who have been worshipped by some and feared as demons by others in India since ancient times. It is said to be a massive, seven-level complex of huge caverns and tunnels, deep underground. The serpent people are believed to reside there mainly in their capital city of Bhogawati. It is

Mountains surrounding Lake Manasarovar in Tibet

known that there are at least two entrances to the world of the Nagas. One entrance is at Sheshna's Well in Benares, and the other is in the mountains around the beautiful Lake Manasarovar about five hundred miles west of Lhasa. At fifteen thousand feet elevation, it is the highest freshwater lake in the world, and is said to have been favored by the Buddha as a meditative retreat. Bruce Alan Walton, also known as "Branton" (now deceased), has emerged on the Internet as one of the most authoritative figures about alien colonies on Earth. He claims that locals around the lake have reportedly seen the Reptilians in that region and have seen their wingless flying craft entering and leaving the mountains. We know now that the Reptilians are closely associated with the so-called Greys originally from Zeta Reticuli, so it would be very likely that a Grey colony also exists in Patala.

THE NAZI GODFATHER

In the years between the armistice and 1933, Karl Haushofer actively sought the man who would lead Germany and make it into a fascist military power, capable of joining with Japan to take over Russia, either by alliance or conquest, and thus rule the entire Eurasian landmass. Already well respected and prestigious, he became an associate professor of geopolitics at Munich University in 1919, and was thus uniquely positioned to advise the chosen leader about this Green Dragon territorial agenda. It seems highly unlikely that, as a humble academic, he would have chosen to undertake such an ambitious political role if he were not acting as an agent for the Japanese secret societies. And it has become abundantly clear from the results of this endeavor that he was actively assisted and advised by his Black and Green Dragon mentors. In preparation for designating the new German dictator, Haushofer was instrumental in founding two secret organizations with ties to the Dragons. Along with Rudolf von Sebottendorf, he helped found the occult-based Thule Society in Munich in 1918. At its height, this organization had about fifteen hundred members in Bavaria, many of whom were affluent and influential in German industry and right-wing politics. The Thulists eventually morphed into the Nazi Party. So Haushofer had thereby set the stage for support of the new leader. He also founded the Vril Society, which had ties to the Tibetan monks. It was in this super-secret inner core of the Thule that the first German flying discs were developed. Clearly, the antigravity technology was being funneled from Patala, where the Reptilians were known to be using these craft, to the Vril Society through the Green Men. Haushofer arranged to bring a group of monks from the Green Dragons and the Society of Green Men to Berlin to set up a scientific advisory group.*

Encouraged by Rudolf Hess, Haushofer attended Adolf Hitler's trial for treason in Munich in 1923. Hitler so impressed Haushofer, with his electrifying oration at the trial, that he decided that Hitler was the chosen one. Through the mediation of Hess, who was close to Hitler, Haushofer began Hitler's indoctrination and education in his jail cell at Landsberg Prison. He visited Hitler there frequently in 1924, and wrote all the sections about geopolitics in *Mein Kampf.* Then, through his Thule connections, Haushofer was able to influence German industrialists to finance the rise to power of both Hitler and the Nazi Party. Haushofer, as founder of the Institute of Geopolitics in Munich in 1922, began organizing annual trips to Tibet for his students and followers as early as 1926, evidently linking up with his Dragon connections on these trips. When Hitler came to power in 1933, Haushofer arranged for a pact between Hitler and the Reptilians from Patala through the mediation of the Green Men. At that point, the Vril Society became the technical arm of the SS, and the antigravity disc development became an SS operation.

Without question, Karl Ernst Haushofer was the godfather of Nazi Germany. It was he who convinced Hitler that the Germans were a master race descended from the Aryan survivors of the Atlantean deluge. It was he who invoked the term Lebensraum (living space) to justify Germany's taking over the lands of adjacent, "inferior" countries without conscience. Japan, with its world-class imperial navy, would rule the seas to protect those territorial gains. The Green Dragons even

*Other advanced German weaponry was also developed through scientific information funneled from Patala through the Green Men in Berlin, the so-called Wunderwaffe or "wonder weapons."

supplied Hitler with a core army of one million cloned warriors, the fearsome, intrepid Wehrmacht.* It seemed a foolproof plan. All the details had been meticulously worked out by the Black Dragons, and Karl Haushofer carried out his mission to perfection. However, there was one unpredictable element, and it proved to be the undoing of the entire operation. They were not able to control Adolf Hitler, who simply refused to be a good little puppet, and ultimately turned out to be a madman. When he insisted on becoming the military commander in chief, and undertook to fight two powerful adversaries on two fronts while, at the same time, carrying out an insane, massive genocidal program, a Götterdämmerung (Twilight of the Gods), ending became inevitable. Eventually, Hitler turned on his mentor and sent Haushofer and his family to a concentration camp. In the end, with Germany in ruins, Haushofer, although ironically already absolved of blame by the Nuremberg tribunal, realized that he had picked the wrong man, and he ended his life in the only correct way for a failed Green Dragon. He and his wife committed suicide in early 1946. As Westerners, they did not feel obliged to use the brutal Japanese Hara-Kiri method of a stomach slash, but took poison instead. His wife also hanged herself, evidently just to make certain that she would die.

THE GERMAN DISCS

In 1944, SS Commander Heinrich Himmler withdrew all secret technology and weapons development from under the control of Hermann Goering and relegated it to civil engineer SS General Hans Kammler, and it was removed to the massive Skoda Munitions Works near Pilsen, Czechoslovakia. By this action, Kammler became the third most powerful man in Nazi Germany. The Skoda factory produced the German Panzer tanks early in the war and had the capacity for large-scale metal castings, which were needed to build the discs. The scientific and technological foundations for the Nazi disc development came from Patala and

*See *The Secret History of Extraterrestrials* by this author, chapter 23 (Inner Traditions, 2010).

was supplied to the SS scientists by the Society of Green Men who had set up an advisory colony in Berlin. There is evidence that the Germans manufactured up to twenty-five working models of the Haunebu type. This was the distinctive bell-shaped craft powered by a rather simple electrogravitation motor called the Kohler converter, developed by Captain Hans Kohler based on the Tesla coil. This motor converted the Earth's gravitational energy into electromagnetic power, but could also extract energy from the ambient vacuum in outer space. In this series, the *Haunebu I* was a small, two-man ship, but the *Haunebu II* was much larger and more sophisticated. It was reported to have a diameter of about seventy-five feet, and had the capacity to carry a full crew. The German SS plans for this craft, dated November 7, 1943, are available on the Internet. Also available are photos of the Haunebu in flight, clearly showing the German cross painted on the side and flanges, and a 7.5mm antitank gun mounted on a swivel turret, apparently identical to the gun then being used on German Panzer tanks.*

Other important antigravity weapons research was carried on near Prague, presumably at Skoda, primarily by Viktor Schauberger and Richard Miethe. Miethe, in cooperation with the Italians, developed the large helium-powered V-7 and the small one-man Vril models, which achieved a speed of 2,900 km/hr in flight tests. In a letter written by Schauberger to a friend, he gives information from his direct experience. He says:

> The flying saucer which was flight-tested on the 19th of February 1945 near Prague and which attained a height of 15,000 meters in three minutes and a horizontal speed of 2,200 km/hr, was constructed according to a Model I built at Mauthausen concentration camp in collaboration with the first-class engineers and stress analysts assigned to me from the prisoners there. It was only after the end of the war that I came to hear . . . that further intensive development was in progress . . . at the works in Prague.

*For photos, illustrations, and complete information, see http://discaircraft.greyfalcon.us/HAUNEBU.htm.

Nazi General Hans Kammler

In *German Secret Weapons of the Second World War,* Rudolf Lusar says, "The development . . . which had cost millions, was almost complete by the end of the war." Noted Hungarian physicist-researcher Vladimir Terziski says that by this time the German technicians had built a huge version of the *Haunebu,* which was about 230 feet in diameter. This "dreadnought," the *Haunebu IV* (see plate 1), was piloted by an all-volunteer, joint German-Japanese crew and sent on a "suicide" mission to Mars. According to Terziski, it crash-landed on Mars in January 1946 after a difficult eight-month flight. This means that it departed the Earth right around the time of Hitler's suicide and the German capitulation, which means that it could not have left from Germany. Terziski says that this mission left from the joint Nazi-alien base at Neuschwabenland in Antarctica (see next chapter). The Green Men died in the final days as Berlin was virtually destroyed by the bombings and advancing Allied armies. Their bodies were discovered by the Russians in the rubble, arranged in a circle. They were wearing German uniforms.

In April 1945, a French diplomat living in Switzerland wrote the following report:

> The circular German fighter without wings or rudder suddenly overtook the four-engined Liberator, crossing its flight path at very high speed. When it passed in front of the formation, it gave off a number of little bluish clouds of smoke. A moment later the

American bombers mysteriously caught fire, exploding in the air, while the German rocket had already disappeared over the horizon.

The circular fighter was the final product of the long years of Nazi research and experimentation in eight different areas: direct gyroscopic stabilization; television-controlled flight; vertical takeoff and landing; jam-free radio control combined with radar blinding; infrared search eyes; electrostatic weapon firing; hypercombustible gas combined with a total reaction turbine; and antigravityflight technology. This was the incredible Kugelblitz, or "lightning ball." If it had emerged even six months earlier, the war could have turned out very differently. It was the last gasp of the Third Reich, but it was an ominous harbinger of things to come.

KAMMLER DISAPPEARS

In mid-April 1945, as General George S. Patton's Third Army rapidly approached Berlin on a direct eastern trajectory, General Eisenhower commanded him to halt and change direction. He was sent southeast toward Prague, Czechoslovakia, and then was told to stop at Pilsen, the home of the Skoda works. Patton obeyed these orders very reluctantly, since he had been intent on beating the Russians to Berlin. Evidently, Eisenhower had been informed by the OSS of the secret weapons development there under Kammler. Patton arrived at Skoda six days before the Russians, but Kammler was gone. On February 23, 1945, the newest engine of the Kugelblitz had been shipped out, and the shell was blown up. Two days later the underground plant at Kahla, Germany, was closed and all the slave workers were sent to Buchenwald to be gassed and cremated in accordance with the grisly Nazi credo, "Dead men tell no tales." Kammler was in charge of the evacuation. He was never found. The various intelligence agencies of the Allied powers had gotten all the espionage reports, and were well aware of what was going on in Hitler's underground Alpine facilities. Consequently, the invading armies knew precisely what to look for. According to Italian author Renato Vesco in his book *Intercept UFO,* "tons of blueprints, company papers, lists of researchers, laboratory models, memoranda, reports, and notes that covered every sector of the war industry came pouring out of thousands of unlikely hiding places." Most certainly, we can conclude from this that much of the antigravity information fell into Allied hands. At that point, it is very unlikely that the Allied powers understood where the antigravity technology had originated. The discovery of the bodies of the Green Men in Berlin might have offered a clue, but there was no other evidence that it came from Tibet. Undoubtedly, they believed that it had been developed by the German scientists.

DISC TECHNOLOGY AT WRIGHT-PATTERSON

Viktor Schauberger, the Austrian scientist who had designed the flying disc built at the Mauthausen concentration camp, was apprehended by U.S. intelligence agents immediately after the war, and was kept in their custody for nine months. The agents confiscated all his documents, notes, and prototypes, and interrogated him intensively. He was then sent to the United States to continue to work on his innovative antigravity disc.

Oddly enough, perhaps coincidentally, Dr. Eric Wang, a fellow Viennese aeronautical engineer, was then teaching at the University of Cincinnati. Wang had received his engineering degree at the Technical University of Vienna in 1935. Little or nothing is known about his movements after that until, in 1943, he was on the staff of the university where he taught engineering and mathematics. Presumably, he had emigrated to the United States before the war, as did many other German and Austrian scientists, including Albert Einstein. In 1949, he was recruited by the Air Force to work at Wright-Patterson AFB in the Office of Foreign Technology. This is where the crashed alien discs from New Mexico were brought for analysis and reverse engineering. Wang is known to have said that the flying disc technology he was working on for the Air Force was different than the Schauberger technology. We can reasonably conclude from that remark that Wang was working with Schauberger under Air Force auspices when Schauberger was sent to the United States, even before Wang officially went to work

Viktor Schauberger

for the Air Force. That would have been between 1945 and 1949. It is known that Viktor Schauberger later joined research efforts in flying disc development in Texas. It is believed that the original disc itself was destroyed by the Germans, along with prototypes of the Schriever-Habermohl-Miethe model. This was the legendary V-7. It is known that Klaus Habermohl was taken to the Soviet Union and some think that the Russians succeeded in obtaining a prototype of the V-7 when they arrived at Skoda. Miethe went to work for the United States and Canada. So we can conclude from this that the Army Air Force (the precursor to the Air Force) knew all about electromagnetic antigravity disc technology, probably as early as 1944. However, as will be seen in the next chapter, in early 1947 the military learned just how deadly the discs could be in actual combat.

2
ANTARCTICA

Prior to the 1930s, the continent of Antarctica drew scant attention as a possible place to establish a permanent colony. There really was no reason why any civilized nation would want to consider such a frozen, forbidding land as a place for human habitation. But it did become a magnet for intrepid explorers who wanted to make noteworthy discoveries, or to be the first to reach the South Pole. Credit for the discovery of Antarctica is given to Russian naval officer Fabian Bellingshausen, who first sighted land in the Southern Ocean on January 28, 1820, and circumnavigated the continent twice. There were several expeditions by Europeans in the nineteenth century, notably by the British, the Belgians, and the Norwegians. From 1839 to 1843, daring British naval officer James Clark Ross charted much of the coastline and discovered and named the Ross Sea, Victoria Land, and the volcanoes Mt. Erebus and Mt. Terror, both named after expedition vessels. Ross was knighted upon his return to England, and also received the French Legion d'Honneur.

THE NAZI BASE

With the advent of aviation, reaching the South Pole by air became a real possibility, and it was accomplished by seasoned aviator Richard Evelyn Byrd on November 28, 1929, earning him the gold medal of the American Geographical Society. Byrd's expedition established a base camp on the Ross Ice Shelf, called "Little America," and commenced exploration of the continent by snowshoe, snowmobile, dogsled, and airplane. Byrd's second Antarctic expedition in 1934 nearly ended in tragedy. He spent five winter months all by himself in a small meteorological advance station where he was overcome by carbon monoxide from a small heater, but he was rescued in time by team members from the base camp. This harrowing adventure was described in his book, *Alone,* first published in 1938.

Admiral Richard E. Byrd

Given this sparse international history of the rough-hewn exploration of the continent, it was surprising, even astonishing, that the Germans would seek to establish a colony in Antarctica in 1938. There had been two previous German expeditions, in 1901 and 1911, each lasting two years, but those pre-Nazi journeys offered no hint that the Germans actually wanted to live there. But the Nazis were very serious about this project. The preparations for the German Antarctic

Germany's 1938 Antarctic Mission Emblem

expedition of 1938 were massive and comprehensive. The Nazis even brought Richard Byrd, the preeminent world authority on Antarctica, to Hamburg prior to the mission departure to advise the members of the team. They also asked him to accompany the expedition, an invitation he declined. Byrd was a civilian at the time, and his agreement to counsel the team members did not, in any way, constitute sympathy with the Nazi regime. Byrd, of course, had to be aware of the German expansionist intentions, since Hitler had already taken over Austria. But after the Munich Agreement in September 1938, the world lapsed into the delusion that Hitler had no further territorial ambitions. Then there was the distinct possibility that Byrd went there on a U.S. government mission to covertly obtain information about Germany's plans for Antarctica.

The Germans utilized the seaplane carrier *Schwabenland* for the 1938 journey. According to Russian ufologist Konstantin Ivanenko:

> The *Schwabenland* sailed to Antarctica, commanded by Albert Richter [Ritscher], a veteran of cold-weather operations. The Richter [Ritscher] expedition's scientists used their large Dornier seaplanes to explore the polar wastes, emulating Admiral Richard E. Byrd's efforts a decade earlier. The German scientists discovered ice-free lakes (heated by underground volcanic features) and were able to

land on them.* It is widely believed that the Schwabenland's expedition was aimed at scouting out a secret base of operations.

The seaplanes dropped swastika flag pennons all over the Queen Maud Land area, staking out a widespread German territory amounting to 600,000 square kilometers (about 360,000 square miles). They then established a base in the Muhlig-Hofmann Mountains, very close to the Princess Astrid coast, which they referred to as Neuschwabenland, named after the duchy of Swabia, part of the original German kingdom.

NEUSCHWABENLAND

The legend and lore surrounding the Nazi Antarctic base is voluminous, bordering on overwhelming. Various writers say that German convoys started bringing in equipment to develop this base beginning in 1938. According to the Omega File:

> Beginning in 1938 . . . the Nazi's [sic] commenced to send out numerous exploratory missions to the Queen Maud region of Antarctica. A steady stream of expeditions were reportedly sent out from [at the time] white supremacist South Africa. Over 230,000 square miles of the frozen continent were mapped from the air, and the Germans discovered vast regions that were surprisingly free of ice, as well as warm water lakes and cave inlets. One vast ice cave within the glacier was reportedly found to extend 30 miles to a large hot-water geothermal lake deep below.

In October 1939, one month after the beginning of World War II, the *Schwabenland* was turned over to the Luftwaffe, which means that Hermann Goering took over the project. On December 17, 1939, the ship again left Hamburg headed for Antarctica, loaded with scientists and equipment. This time they were going to build a permanent base. The Omega File says:

*The warm-water lakes are known as the "Rainbow Lakes" because of the colors of the algae.

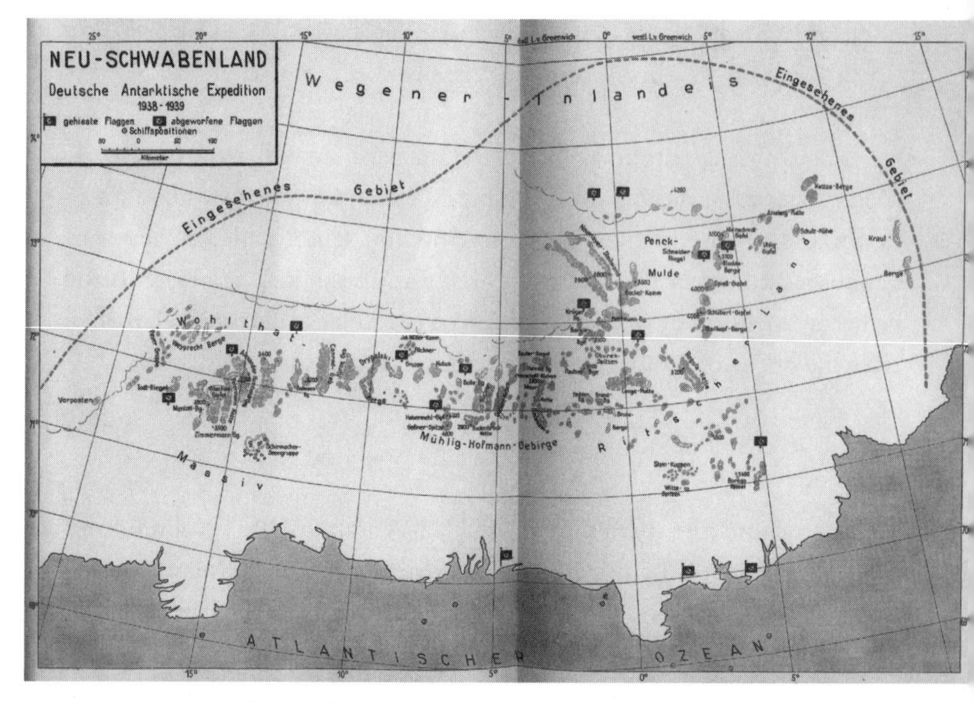

Map of Neuschwabenland; flags show German colonies

Various scientific teams were moved in to the area, including hunters, trappers, collectors and zoologists, botanists, agriculturists, plant specialists, mycologists, parasitologists, marine biologists, ornithologists, and many others. Numerous divisions of the German government were involved in the top secret project.

The development effort continued throughout the war. An article appearing in *The Plain Truth* magazine (U.S.) in June 1952 says, ". . . in 1940 the Nazis started to amass tractors, planes, sledges, gliders and all sorts of machinery and materials in the south polar regions" and ". . . for the next 4 years Nazi technicians built, on . . . Antarctica, the Fuhrer's Shangri-La." The article goes on to say, ". . . they scooped out an entire mountain, built a new refuge completely camouflaged—a magic mountain hideaway." According to the original *Bonjour* magazine, a Parisian publication in the 1950s, the Nazi engineers constructed buildings at this base in 1940 that could withstand temperatures up to –60° Fahrenheit. There are many cave inlets in this region, allowing submarine access to a relatively comfortable disembarkation. Engineer and physicist Vladimir Terziski says that the Germans populated a city called Neu Berlin under the ice with a community of scientists and workers that approximated forty thousand civilians by the end of the war. This city was just a small section of the vast Neuschwabenland colony under Queen Maud Land. Presumably, the population was supplied with food via merchant vessels from Argentina and by growing crops hydroponically. Then, of course, the Southern Ocean is rich with fish and marine animals. It is also possible that the Germans had begun agricultural projects in the warm-water lakes region of Antarctica.

THE RUINS OF KADATH

Considering that Germany was only one year away from embarking on the invasion of Poland, and thereby starting World War II, it is astounding that the Germans would have devoted such energy and so many resources to the development of a base on a frozen wasteland five thousand miles from Germany. Several other sources have claimed that Antarctica was actually previously Atlantis, which moved to the South Polar region as a result of a prehistoric pole shift.* Since the Reptilians are known to have inhabited Atlantis, it is very possible that their colony remained underground after the shift and they were still living under Antarctica. Ivanenko says that Neu Berlin adjoins "the prehistoric ruins of Kadath [à la the novella by H. P. Lovecraft], which may have been built by settlers from the lost continent of Atlantis well over 100,000 years ago." Another unnamed occult researcher claimed that "Neu Berlin has an alien quarter where Pleiadians, Zeta Reticulans, Reptoids, Men in Black, Aldeberani and other visitors from the stars dwell."

LUNATIC LEADERS

It is not at all far-fetched to conclude from this that the Reptilians encouraged the Nazis to establish a colony adjoining theirs, possibly as

---

*See *Atlantis beneath the Ice: The Fate of the Lost Continent* by Rand Flem-Ath (Bear & Co., 2012).

a refuge in the event the war turned out badly. But more likely, this colony was to be a base for joint German-alien scientific and technological development for interplanetary travel and conquest.* As we have already seen, the Nazis had executed a pact with the Reptilians living in Patala in 1933 that resulted in a transfer of advanced technology to Germany, including antigravity discs. This would explain why, as we will see below, the Nazis eventually moved all the aeronautical engineers and scientists involved in the development of the antigravity discs, as well as the prototypes themselves, to Neuschwabenland. At that early date, the Nazis were supremely confident of winning the war, and would not have been thinking about a refuge. It would, therefore, be reasonable to conclude that this is the real reason—and the only reason—that the Nazis went to such trouble to build not just a base, but a city, under two miles of ice halfway around the globe. After all, they envisioned the conquest of the entire Eurasian landmass, so it's not as though they were so concerned about territorial deprivation that they would have to move to the South Pole to survive as a nation.

It appears that, even at this early stage, the Nazis were already anticipating travel to other planets alongside their Reptilian compatriots. It was from this Antarctic base that the joint German-Japanese suicide mission to Mars was launched in mid-1945 (see chapter 1). The German population knew nothing about any of this. It is certainly understandable that the Nazi leadership would not have wanted the German people to know about all these fantastic plans. They would surely have concluded that their leaders were raving lunatics! But perhaps the most important reason for the secrecy was the fact that the Nazi overlords had no intention of relocating the entire population of Germany to Neuschwabenland. This colony was to be reserved for only the purest specimens of the Aryan race.

---

*According to "Branton," the author of *The Omega File,* the Draco Reptilians, in alliance with extraterrestrials from Orion, had already conquered and enslaved civilizations in twenty-one star systems in a nearby section of the galaxy. Possibly, they had chosen the South Polar colony as a jumping-off point for their missions in this galactic neighborhood, probably because it was remote and unpopulated. It is becoming increasingly clear to researchers that Hitler was allied with extraterrestrials from the very beginning, which helps to explain why he believed that he was invincible.

ANTARCTIC SETTLEMENT WOMEN

By mid-1943, the Allies noticed heavy submarine traffic in the South Atlantic, and began to suspect that something unusual was going on. The Neuschwabenland base construction and settlement had been removed from the control of failed Luftwaffe chief Hermann Goering and was now under the control of Heinrich Himmler, who was using the large Milchkuh (milk cow) supply submarines to transport personnel and equipment to the Antarctic colony. These special XXI U-boats, diverted from Atlantic warfare, were almost as large as tramp steamers, and, thanks to a newly developed snorkel, were capable of making the entire journey underwater. According to Ivanenko, Himmler selected ten thousand of the "racially most pure" Ukrainian women out of the half-million ethnic German women who were deported from Russia, and sent them to Neu Berlin on the submarines. They were all blond and blue-eyed and between the ages of seventeen and twenty-four. These were Himmler's Antarktisches Siedlungensfrauen, or "Antarctic settlement women." He also sent 2,500 battle-hardened Waffen-SS soldiers who had been fighting on the Russian front. At four women to each soldier, they were expected to breed the Aryan population of the new civilization under the Antarctic ice.

Since the European war appeared lost in 1944, the huge submarines also carried the prototypes of the antigravity discs, along with the crucial aeronautical engineers and scientists (see plate 2). It is very likely that Hans Kammler, who reported to Himmler, and all his important workers and technicians, along with the designs and raw materials for the new antigravity aircraft, were taken to Neuschwabenland in 1945. And there is compelling evidence that Hitler himself may have been taken there while his double was executed and made to look like a suicide, in the underground Berlin bunker.* All during that period, German warships patrolled the South

---

*At the Potsdam conference in July 1945, Soviet leader Joseph Stalin told U.S. President Harry S. Truman and British Prime Minister Winston Churchill that Hitler was still alive. He is reported to have said, "Hitler had escaped and no traces of him were found."

Atlantic waters and sank merchant vessels that they encountered anywhere near Tierra del Fuego in Argentina, the gateway to Antarctica. The deadly pocket battleship *Graf Spee* was stationed in the South Atlantic and sank ten merchant vessels from various countries in the early days of the war. It would appear that the Germans were concerned that reports about the increased submarine traffic would be relayed to the Allies, which might imperil the Antarctic settlement.

According to noted Third Reich researcher/writer Rob Arndt, writing on his website, antarctica.greyfalcon.us, after the war, the Allies were able to determine that fifty-four U-boats were missing from Nazi Germany. He says also that between 142,000 and 250,000 people were unaccounted for, including the entire SS Technical Branch, the entire Vril and Thule Gesellschafts, 6,000 scientists and technicians and tens of thousands of slave laborers.* He claims that this information was obtained from high-level declassified communications between Washington and London in late 1945 and 1946. This conforms to other information that Neuschwabenland was primarily a technological and scientific development colony, meant to be inhabited by only a very small, select "pure Aryan" group. The rest of the German population was left to perish in the Allied onslaught. This core Aryan colony would then breed the new race. With their super weapons and their extraterrestrial friends, the Neuschwabenland civilization would then be positioned to establish the Fourth Reich to take over and enslave the "inferior" races on the rest of the planet. This planetary conquest would be facilitated by the expected next 90-degree pole shift, which would eliminate most of the existing population, and would move Antarctica back to a temperate climate near the equator. They would remain safe throughout the catastrophic events in their secure redoubt under the two-mile mantle of ice.

THREE U-BOATS SURFACE

According to UFO researcher/writer Erich J. Choron, ten of the missing German U-boats participated in a top secret mission in the last days

*Very probably, the slave workers were not brought to Neuschwabenland, but were gassed and cremated at Buchenwald to ensure silence about the advanced weaponry.

of the war. In an article titled "How High Can You Jump?" in *The UFO Casebook* (vol. 26, no. 4), Choron says:

> The fact that in the dying moments of the Second World War, ten U-boats, based in Oslofjord, Hamburg and Flensburg, were made available to transport several hundred German officers and officials to Argentina to found a new Reich is widely accepted. These officers, mostly involved in secret projects, and many of whom were members of the SS and Kriegsmarine, itself, sought to escape the "vengeance of the Allies" and continue their work, abroad. The U-boats were filled with their luggage, documents and, more than likely, gold bullion, to finance their efforts . . . Seven of the ten of the U-boats, based on the German/Danish border, set off for Argentina through the Kattegat and the Skagerrak. None were ever seen again . . . "officially."

From what we have already seen, it seems evident that these submarines went to Antarctica, which was to be the home of the Fourth Reich, not Argentina. Argentina had joined the Allies in March 1945 and was now hostile to Germany. Choron also says that many of the missing U-boats were the very advanced Type XXI and Type XXIII, manufactured late in the war, able to travel much faster than previous models, and equipped with a new snorkel that allowed them to make the entire transatlantic voyage underwater. These could easily evade Allied warships in the South Atlantic.

All these submarines were known to have departed their home ports between May 3 and May 8, 1945. The naval war ended on May 5, 1945, when Admiral Karl Doenitz commanded all submarines to surrender, although the official German surrender was on May 8th. Three of these ships did eventually show up. U-530, under the command of Oberleutnant Otto Wermuth, and U-977, commanded by Oberleutnant Heinz Schaeffer, surrendered to the Argentine Navy at Mar del Plata on July 10, 1945, and August 17, 1945, respectively. U-1238 was scuttled by her crew off northern Patagonia, which is at the tip of South America, and was very likely en route to or from Antarctica. Wermuth and Schaeffer were intensely interrogated by both the United States and Great Britain before being released as civilians. It is very likely that information gathered in these interrogations triggered both Operation Taberin and Operation Highjump, although the British already had incriminating information of their own that they had not shared with their U.S. counterparts.

OPERATION TABERIN

According to James Robert, a British civil servant and World War II historian, in an article in the August 2005 edition of *Nexus Magazine* (vol. 12, no. 5), the Germans succeeded in building an underground base in the massive ice cave, using the discovered inlets for access. He claims that British soldiers from the secret Antarctic Maudheim Base found the entrance in late 1945 and

> followed the tunnel for miles, and eventually they came to a vast underground cavern that was abnormally warm; some of the scientists believed that it was warmed geothermally. In the huge cavern were underground lakes; however, the mystery deepened, as the cavern was lit artificially. The cavern proved so extensive that they had to split up, and that was when the real discoveries were made. The Nazis had constructed a huge base into the caverns and had even built docks for U-boats, and one was identified supposedly. Still, the deeper they traveled, the more strange visions they were greeted with. The survivor reported that "hangars for strange planes and excavations galore" had been documented.

The Nexus article included a long-kept secret, firsthand account of what was called Operation Taberin II* in October 1945, by a former British SAS (Special Air Service) commando, who had partici-

*Operation Tabarin I was carried out in 1943.

pated in the raid and survived.* While Britain had several secret bases in and around Antarctica, the Maudheim base was the largest and was top secret because it was only about two hundred miles from the Muhlig-Hofmann Mountains, and it was from Maudheim that the attack was launched. During training for the operation, the agent was informed that Antarctica was Britain's secret war. The British bases in Antarctica had been set up in anticipation of an eventual confrontation, after they had learned about the construction of the Nazi base in 1939. This information was revealed by three key Nazis who were captured by the British—Rudolph Hess, Heinrich Himmler, and Admiral Karl Doenitz, all of whom knew all the details of the secret base—and by submarine commanders Otto Wermuth and Heinz Schaeffer. In fact, it was highly probable that Doenitz was named as Hitler's successor precisely because, as commander of the submarine fleet, he was best positioned to protect the Antarctic colony, the future home of the Fourth Reich. This choice by Hitler came as a great surprise to the entire German High Command.

After a month of arduous cold-weather training, the special commando team was informed that the tunnel to the Nazi base had been discovered during the previous Antarctic summer and explored by a previous SAS team. Of that team of thirty men from the Maudheim base, there was only one survivor who had succeeded in somehow lasting through the Antarctic winter without going stir-crazy. He told the new team what they had discovered and how the others had died. The new team set up an advance base at the mouth of the tunnel, and then were told they had orders to follow the tunnel all the way "to the Führer, if need be." Two men remained behind with the radio and other equipment while eight commandos, led by a major and carrying a huge amount of explosives, went into the tunnel. After walking for

*The SAS was an elite British-Canadian commando organization, primarily paratroopers. These were highly trained warriors, similar to Navy SEALs, who were called upon for special operations. For most missions they parachuted behind enemy lines. They played an important role in World War II, harassing German installations from behind the front lines using heavily armed Jeeps and light artillery also parachuted down behind the lines.

Daily Mail 4 A.M. EDITION

NO. 15,285 ONE PENNY FOR KING AND EMPIRE WEDNESDAY, MAY 2, 1945

The most dramatic news of the war

'HITLER DEAD—DOENITZ APPOINTED FÜHRER'

Admiral tells Germans: 'The fight goes on.' Himmler ignored

ADOLF HITLER, is dead. Grand Admiral Doenitz, Commander-in-Chief of the German Navy, has been appointed the new Führer. The German radio gave the news to the world at 10.25 last night in the following words: "It is reported from the Führer's headquarters that our Führer, Adolf Hitler, has fallen this afternoon in his command post in the Reich Chancellery fighting to his last breath against Bolshevism.

"On April 30 the Fuhrer appointed Grand Admiral Doenitz as his successor. The Grand Admiral will now speak to the German people."

Admiral Doenitz, who immediately came on the air, said his task was to save the German people from annihilation at the hands of Bolshevism. For that aim only, the war would go on. Germany would have to continue the fight against Britain and America as long as they "hindered this purpose".

The naming of Doenitz as the new Führer comes as a bombshell. It suggests that what remains of Germany has been split into two camps — those who wish to fight on, led by Doenitz, and those who want to surrender, led by Himmler.

It is significant no reference was made in the announcement to Himmler who has already offered unconditional surrender to Britain and the United States, and was expected to comply with the Allied demand that capitulation must be made also to Russia.

None of the big names of Nazism were there to pledge support to the new Führer. Where was Goebbels? Where was Ribbentrop? No marshal or general was called to the microphone to give force to an Order of the Day by Doenitz calling on the Army to fight on.

'WE FIGHT SOVIETS'

Doenitz said: "German men and women, soldiers of the German Army, our Führer, Adolf Hitler, has fallen. The German people are bowed in sorrow and reverence.

"Our Führer had recognised very early the grim danger of Bolshevism and consecrated his life to the struggle against it.

"At the end of his struggle he met a hero's death in the capital of the German Reich.

Doenitz cannot hold Reich together

By WILSON BROADBENT
Diplomatic Correspondent

THE Prime Minister will make a full statement in the House of Commons today about the events in Germany and the new situation which has arisen following Hitler's death.

THE HATER OF BRITAIN TAKES OVER

Food ships into Holland, and—

Surrender begins on three fronts

REPORTS received in London late last night indicated that large German forces on widely separated fronts have begun to surrender piecemeal to the Allies with or without authority from Doenitz or Himmler.

DENMARK. — German occupation forces were reported from Stockholm to be evacuating the country with all speed. King Christian and the Danish Royal Family are all back in Amalienborg Castle, Copenhagen.

NORWAY. — Negotiations were said to be going on for the German garrisons to lay down their arms at the Swedish frontier to escape Norwegian reprisals.

CZECHO-SLOVAKIA. — A delegation of German and Czech industrialists was reported by Luxemburg radio to have left Prague to meet Allied representatives and hand over the territories of Bohemia and Moravia.

SOUTHERN GERMANY. — Goerlitz radio announced that an important announcement would be made by the Gauleiter of the Upper Danube territory this morning.

ITALY. — Marshal Graziani and Lieut. General Pemsel, German Chief of Staff of the Italian Fascist Ligurian Army, last night announced the surrender of that army.

HOLLAND. — Food ships for the starving Dutch are to enter Rotterdam almost at once. Food is also to enter the German occupied area of Holland by road starting to-day.

'IMPORTANT NEWS TO-DAY' — GERMANS

Görlitz (South Germany) Radio announces: "Important news will be broadcast between 10.30am and noon to-day." — A.P.

'TRICKERY'- MOSCOW

Moscow radio said German announcement of Hitler's death "repeats the usual trickery and twists of Hitlerite propaganda." — A.P.

British newspaper report of the Doenitz appointment

five hours, they entered an enormous cavern illuminated by artificial lighting. In his account, the SAS agent says, "As we looked over the entire cavern network, we were overwhelmed by the numbers of personnel scurrying about like ants, but what was impressive was the huge constructions that were being built. From what we were witnessing, the Nazis, it appeared, had been on Antarctica a long time." He says he was very impressed by the advanced Nazi technology. The team was discovered and fought a heroic engagement while being chased after setting the mines in place. Only three survived the encounter, but they succeeded in detonating massive explosions at the mouth of the tunnel, and sealing it so that no entrance remained. After being evacuated to the Falkland Islands, the three survivors were told that their mission was to remain top secret. The SAS agent says, "Upon reaching South

Georgia, we were issued with . . . a directive that we were forbidden to reveal what we had seen, heard or even encountered."

OPERATION HIGHJUMP

Clearly, the United States did find out about Operation Tabarin, either by covert intelligence operations or by deliberate intelligence sharing by the British. The latter seems more likely because the British were probably convinced they had not succeeded in destroying the base, and wanted the U.S. to finish the job. Also, the OSS had learned a great deal from the interrogations of Wermuth and Schaeffer. Planning for Operation Highjump was initiated by Secretary of the Navy James V. Forrestal on August 7, 1946, less than a year after Operation Tabarin II concluded. Operation Highjump was sanctioned by order of the "Committee of Three," consisting of the secretaries of state, war, and the navy. Presumably, the Cabinet had been advised by several intelligence agencies and already had the approval of President Harry S. Truman. This was to be a massive naval operation involving a fleet of thirteen ships, including a communications-laden flagship, two icebreakers, two destroyers, two tenders carrying three PBM (Patrol Bomber Mariner) seaplanes each, two tankers, two supply ships, one submarine, two helicopters, and the aircraft carrier USS *Philippine Sea,* carrying six DC-3 twin-engine planes equipped with both wheels and skis for landing gear. The flagship USS *Mt. Olympus* also carried a contingent of 4,700 Marines.

War hero Fleet Admiral Chester W. Nimitz, who was also then U.S. chief of naval operations, appointed Rear Admiral Richard E. Byrd as director of the mission. He also named decorated veteran of polar operations, Admiral Richard H. Cruzen as task force commander. The operation was publicized as exploratory and scientific. Clearly, with the involvement of three of America's top naval brass as well as a Marine fighting unit, this was no scientific expedition. The U.S. Marine Corps at that time was considered the toughest military organization in the world, still including in its ranks many veterans of the brutal Pacific Island campaigns only a year earlier. So this was no novice, token military force intended to accompany a scientific expedition.

The publicity notwithstanding, the military nature of the operation was clearly stated from the top down. The Committee of Three said that the main purpose of the expedition was "consolidating and extending U.S. sovereignty over Antarctic areas, investigating possible base sites, and extending scientific knowledge in general." Admiral Marc Mitscher, commander of the Atlantic Fleet, said that the main objective was to extend U.S. sovereignty "over the largest practicable area of the Antarctic continent." Presumably, the presence of such a formidable fighting force implied that this sovereignty might have to be gained by military action. But that made no sense because no potential enemy was identified. So clearly, that was just a cover story, and the real intent of the mission was to covertly destroy the Nazi base. Lest there be any doubt remaining, Admiral Byrd said, "However the basic objectives were not diplomatic, scientific or economic—they were military." Quite possibly, it was actually Byrd who had initiated the entire operation and had convinced the Cabinet of the necessity of the mission because of what he had learned in Hamburg in 1938. This, added to the reports of Operation Tabarin and the U-boat commander interrogations, were sufficient to launch "the largest Antarctic expedition ever organized."*

THE BATTLE OF THE WEDDELL SEA

The operation, planned jointly by Admirals Nimitz and Byrd, involved a three-pronged approach, very similar to an invasion scenario and typical of a military attack. The central group, consisting of two icebreakers, the aircraft carrier, two cargo ships, the submarine, and the flagship, would reestablish the previous base at Little America III, now to be called Little America IV. The six DC-3s would take off from the carrier in the Southern Ocean and fly over the Ross Ice Shelf to the base where a landing strip would be constructed for the planes. They would then conduct reconnaissance flights over the interior using ground-penetrating radar. When ready for the attack, the eastern and western groups, consisting of a PBM seaplane tender, a tanker, and a destroyer

*See www.south-pole.com.

each, would encircle the continent from two different directions, and would rendezvous in the Weddell Sea off Queen Maud Land. Four DC-3s, carrying explosives—one piloted by Admiral Byrd—would then fly over the South Pole from Little America toward Queen Maud Land, while the PBM seaplanes were launched from the tenders. The PBMs (Flying Boats) carried loads of bombs and were capable of sinking ships from the air. They had sunk ten German U-boats during the war.

Descriptions of the expedition never mention the disposition of the Marines. Most likely they would be divided into two groups and carried onboard the tenders and destroyers from which they would be ready to disembark near the mouth of the tunnel from two different directions. The presence of the destroyers and the PBMs clearly signaled the military nature of the operation. So all three groups were to converge on Queen Maud Land. Presumably, the expedition forces had learned from the British the precise location of the entrance to the tunnel leading to the Nazi base, and they now had more current information from the DC-3 reconnaissance flights.

Western group seaplane tender USS Currituck

One of the PBMs flying off the USS *Currituck* seaplane tender in the western group discovered the ice-free zone and the warm water Rainbow Lakes in central Antarctica. The seaplane landed on one of the lakes and found the water temperature to be about 30°F. The eastern group encountered difficulty. The PBM designated *George One,* flying off the seaplane tender USS *Pine Island,* suddenly exploded in midair. Three seamen died in that incident. There was no official explanation for the explosion. The remaining six men in the crew were rescued thirteen days later by the seaplane *George Two,* having survived on the supplies from the plane in the wreckage. More mysterious was the fate of the tender itself. According to Erich J. Choron, in his article "How High Can You Jump?"(see page 37), "The USS *Pine Island* was struck from the Naval Register, on an unknown date . . . Her title was transferred to the Maritime Administration for lay up in the National Defense Reserve Fleet . . . on an unknown date . . . and . . . the ship's final disposition is unknown. . . . Now, how does one go about 'losing' a major surface ship . . ."

According to the official records of Operation Highjump, all the ships in the eastern group met up off Peter I Island in the Bellingshausen Sea on February 14, 1947, and prepared to sail together around the Antarctic Peninsula to the Weddell Sea. It appears that the eastern and western groups were supposed to rendezvous there just off the coast of Queen Maud Land to execute a joint attack with the DC-3s. There is no further record of activity of any of these ships until March 3, 1947, when the operation was suddenly prematurely terminated, and they were instructed to sail back to Rio de Janeiro. According to a Russian documentary posted on YouTube, the fleet encountered several flying discs that emerged from the water and attacked the ships in a twenty-minute engagement. That must have occurred in the Weddell Sea during that period. Apparently, the discs were protecting the entrance to the tunnel. In the video, the discs can be seen darting over a ship. It was claimed that sixty-eight men were killed in that action. If the Pine Island was indeed sunk, it is very likely that it would have happened during that battle, and many of the sixty-eight dead would probably have been Marines.

Whatever happened in the Weddell Sea, it caused Admiral Byrd to cancel the entire operation on March 3, 1947. This was only two months into the mission, originally planned for six months through the Antarctic summer and fall. Onboard his flagship, the USS *Mount Olympus,* as it briefly docked in Valparaiso, Chile, en route back to Washington, D.C., through the Panama Canal, Byrd granted an interview to Lee Van Atta, a reporter from the Chilean newspaper *El Mercurio* on March 4, 1947. The story based on that interview appeared in the newspaper the next day. Van Atta wrote:

> Admiral Richard Byrd warned today that it is necessary for the United States to adopt protective measures against the possibility of an invasion of the country by hostile aircraft proceeding from the polar regions . . . The admiral said: "It is not my intention to scare anybody, but the bitter reality is that if a new war should come, the United States will be attacked by aircraft flying in from over one or both poles . . . The fantastic speed with which the world is developing"—the admiral declared—"is one of the objective lessons learnt during the Antarctic exploration recently effected . . . I cannot do more than deliver a strong warning to my compatriots in the sense that the time has already passed in which we can take refuge in isolation and rest in the confidence that distance, the oceans and the poles constitute a guarantee of security." The admiral reiterated the need to remain in a state of alert and vigilance and build the last redoubts of defence against an invasion.

A NEW BALL GAME

The capabilities of the Nazi flying discs must have made a deep impression on the admiral for him to have issued such an alarmist warning to the most powerful nation on Earth. Byrd arrived back in Washington on April 14, 1947, and was extensively debriefed by Naval intelligence and other government officials. Reportedly, Byrd flew into a rage while testifying before the president and the Joint Chiefs of Staff, and strongly "suggested" that Antarctica be turned into a thermonuclear test range.

After that display, Byrd was hospitalized and was not permitted to give any more interviews or briefings. Now certain that the Third Reich had survived in Antarctica and had perfected their flying discs, it is likely that alarm, and even panic, prevailed at the Pentagon. The antigravity disc development program at Wright-Patterson had not yet yielded a prototype. If, as Byrd warned, the Nazis decided to invade America, we would be powerless to defend ourselves. Very possibly the president and the military might have been considering Byrd's suggestion to employ the nuclear option. But that could blow a hole in the ozone layer over the South Pole, which would have dire environmental consequences. But less than three months later, while several alternatives were being weighed, as if dropped from heaven, an alien craft crashed into the desert near Roswell Army Air Field in New Mexico, and one alien survived. That changed everything, and it became a whole new ball game.

3

ROSWELL

Roswell was not the first incident. We now know that there were at least two other crashes of alien craft in, or near, the United States prior to July 1947. The U.S. Navy retrieved a disc in the Pacific west of San Diego in 1941. Better known was the spectacular crash in the Plains of St. Augustin, southwest of Socorro, New Mexico, on May 31, 1947. The alien craft was resting on its roof and still smoking when the military arrived. Four aliens were on the ground—three alive and one dead. Bob Shell, the former editor of *Shutterbug* magazine and a military cameraman assigned to film the scene, reported that each of the live aliens was tightly grasping a box and making shrieking noises. He said that they looked like "circus freaks." Two of the three surviving aliens were injured and died within three weeks, at which point the cameraman was called upon to film an autopsy of one of the creatures in Fort Worth, Texas. This ultimately became the famous "Santilli autopsy film." As can be seen in the film, these aliens appeared almost human, although smaller, and had six fingers and toes on their otherwise human-looking hands and feet. The alien craft and bodies were taken to Wright-Patterson Air Force Base near Dayton, Ohio.

So the military already had experience with this type of event before Roswell. Given this experience, we can reasonably conclude that Army procedures for dealing with crashed alien craft were in place before Roswell, and that the Pentagon had established a policy to not reveal these events to the press if they had any military implications. That particular crash location was far enough removed from military installations to not have necessarily aroused suspicion of alien surveillance. But any alien presence at all in New Mexico was suspect, since that entire state was, and still is, the very heart of the military-industrial complex. This craft went down not far from the Trinity site at the north end of the White Sands Proving Ground (changed to "White Sands Missile Range" in the mid-sixties), where American and German scientists worked on rocketry, and where the first atomic bomb had been tested on July 16, 1945. Not much farther north, near Santa Fe, was Los Alamos Laboratories, where our most advanced nuclear scientists continued to develop improved atomic bombs. Not far away, in Albuquerque, was Kirtland Air Force Base, where nuclear weapons delivery systems were developed and tested. And the "Z" Division, or Sandia Base, where scientists and engineers worked feverishly on atomic-related weaponry, was also near Albuquerque. Since that incident remained top secret, we can assume that President Truman and the Pentagon believed it to be defense-related.

But the Roswell event would have generated a serious alarm about a possible alien invasion. After all, this craft appeared to be spying on our most sensitive military installation, the 509th Bombardment Group at Roswell Army Air Field near Roswell, New Mexico. This was the home base of the B-29 squadron that had dropped atomic bombs on two Japanese cities two years previously, and was responsible for any future missions involving atomic weapons. This type of surveillance was very suspicious, and was consistent with a possible planned invasion of our planet by an extraterrestrial civilization. As the victors in World War II, and as the only nation possessing atomic weapons at that time, we would have constituted their greatest obstacle to a planetary takeover. It is estimated that by July 1947, we had less than 50 atomic bombs, although plans were in the works for developing an arsenal of 150. Some of them might have still been in development at Sandia, but most of them probably would have already been delivered to Roswell Army Air Field, ready for deployment. This was a very thin margin of protection, and destruction of the Roswell atomic bomb arsenal would have left us, and therefore the entire planet, basically defenseless against the likely advanced weaponry of the aliens. We were not afraid of any Earthly power, but we couldn't defend against an enemy that we knew nothing about, and that had the propulsion technology to travel in space. The fear was compounded when we learned that the alien ship contained human body parts. With the chronicles of World War II just barely in the history books, we were now faced with a potentially catastrophic confrontation that would have made the Axis struggle seem like child's play by comparison. It was a very scary scenario. The U.S. might have solicited the aid of the Soviet Union to try to quickly develop advanced technology and weaponry to fend off an assault. We would not have hesitated to recruit the top scientific brains of our erstwhile mortal enemies, the Nazis. It would have been a massive effort.

In the introduction to his book, *The Day after Roswell,* coauthored with William J. Birnes, Colonel Philip J. Corso sounds the alarm. He says:

> In those confusing hours after the discovery of the crashed Roswell alien craft, the army determined that in the absence of any other

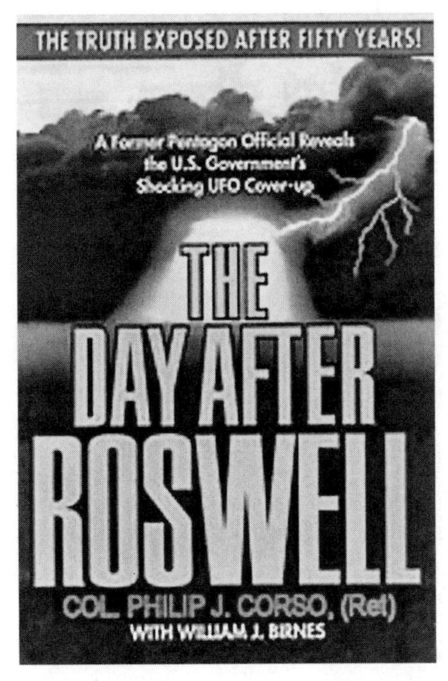

Cover of The Day after Roswell *(Pocket Books, 1997)*

> information it had to be an extraterrestrial. Worse, the fact that this craft and other flying saucers had been surveilling our defensive installations and even seemed to evidence a technology we'd seen evidenced by the Nazis caused the military to assume these flying saucers had hostile intentions and might have even interfered in human events during the war. We didn't know what the inhabitants of these crafts wanted, but we had to assume from their behavior, especially their interventions in the lives of human beings and the reported cattle mutilations, that they could be potential enemies. That meant that we were facing a far superior power with weapons capable of obliterating us.

In 1962, Colonel Corso was given the job of seeding American industry with the objects found in the Roswell disk. He got this assignment from his boss, General Arthur Trudeau. At that point, he had no way of knowing about the events that had transpired in the fifteen intervening years since the crash. That information was so secret and highly compartmentalized that even President Eisenhower didn't have the whole story.

U.S. Army Colonel Philip J. Corso

Only MJ-12, the super-secret advisory committee empaneled by President Truman knew all the details. All Corso knew was that the intriguing pieces of equipment picked up at the crash site displayed technology very far in advance of anything we knew about. His job was to quietly turn that technology over to the scientists and corporations that were best positioned to understand it and to use it as a springboard to perhaps copy it and develop it further. In that role, he was a "lone gunman." He didn't have the clearance to learn about what had happened since 1947.* He just had to assume a civilianlike aspect, and to silently insert these objects into American R&D, and then disappear. So it is not surprising that, when he wrote his book in 1997, he still believed that the aliens known as Ebens were hostile and a potential threat to the United States and the planet. It's an incredible testament to just how effectively the compartmentalization and secrecy machinery functioned that, even fifteen years later, an Army general and colonel had no inkling about the fact that we had been hosting those aliens at Los Alamos Laboratories and Area 51, and had already successfully back-engineered an Eben flying disc!

THE REAGAN BRIEFINGS

President Ronald Reagan was briefed about our experience with crashed alien discs and extraterrestrials between March 6 and 8, 1981, and again between October 9 and 12, 1981. Both briefing sessions took place at the presidential retreat at Camp David, Maryland. The briefings were recorded on fifty-four audiocassette tapes. The tapes were kept in the possession of the Defense Intelligence Agency (DIA) because some of the information discussed a potentially hostile alien race that might have posed a threat to our national security. Consequently, it was appropriate that the cassettes remain in the custody of the Department of Defense. The tapes were declassified in 2007 after the completion of the mandatory twenty-five-year waiting period for the nondisclosure of classified government documents, and they were released into the public domain.

---

*In the manuscript of an unpublished book by Colonel Corso titled *The Dawn of a New Age,* he claims to have clearances up to nine levels above Top Secret. There are believed to be thirty-three levels above Top Secret.

President Ronald Reagan's 1987 UN speech, when he mentioned a possible alien invasion.

The transcript of one of these sessions was published by Victor Martinez as "Release 27a" on the Serpo website, www.serpo.org (e-mail messages from Anonymous to Martinez about Project Serpo were initially referred to as "postings." This convention prevailed for Posting One through Posting Eighteen. After that point, they were called "releases" and were numbered sequentially with numerics by date received beginning with Release 19). This session was conducted during the March 1981 briefing by someone who refers to himself as "the Caretaker."

Martinez precedes the transcript with the following declaration.

> THIS RELEASE HAS BEEN APPROVED BY and AT THE HIGHEST LEVELS OF THE U.S. GOVERNMENT as part of a continuing public acclimation program.

Others present at the briefing were William Casey, the CIA director at that time, and three advisors, as well as a female CIA transcriber. Secretary of Defense Caspar Weinberger and presidential staffer Michael Deaver were present at the beginning, but then bowed out. The following segment of the briefing concerns the Roswell incident, and should be considered to be the most authoritative account of what happened at Roswell.

> **The CARETAKER:** Good morning, Mr. President. First of all, I would like to give you a bit of information on my background. But before that, please, Mr. President, if you have questions during this briefing, just interrupt me, sir. I have been employed by the CIA for the past 31 years. I started the caretaking status of this project in 1960. We have a special group of people whom we call "Group 6," that cares for all this information.
>
> **PRESIDENT:** Good morning, I hope, well, I believe I will ask questions. Bill briefed me back in January, but . . . Bill didn't tell me all 'cause we only had about one hour.
>
> **WM CASEY:** Mr. President, I only gave you a quick briefing for the NSDD (National Security Decision Directives) that we want to incorporate into the overall action directives about this subject. ADVISER #3, Caspar and I have details far beyond what I knew before January. The last administration wasn't too keen on making all of this stuff accessible to us during the briefings in November and December.
>
> **PRESIDENT:** Well, I knew a little about this subject before. Back in 1970. Nixon had all of the good stuff and wanted to share it with some of his friends. Nixon showed me some papers. Not sure about who authored them, but . . . well something about New Mexico and other places. Nixon was pretty . . . fascinated with it. He showed me something, some kind of object or device that came from one of their craft. Something that was taken from the New Mexico crash site. I don't know if, well . . . do we know what it was? I don't think we knew or maybe now, after 11 years, we might know.
>
> **The CARETAKER:** Mr. President, I can answer some of those questions. Do you wish for me to begin?

PRESIDENT: Oh, well, what level is this? I mean, what was it called? I don't remember what they called this.

WM CASEY: Mr. President, codeword. It's called Top Secret Codeword. This information is beyond Top Secret as I said before. It has its own classification. It is very compartmentalized.

PRESIDENT: Well, I guess just the minimum. Are we recording this?

The CARETAKER: Mr. President, no, unless you wish.

WM CASEY: Yes, it is [CIA FEMALE TRANSCRIBER] who is doing that. I think we should. I don't want to make any mistakes later about this. ADVISER #4 should also stay, since he is one of the caretakers of the information.

PRESIDENT: Well, I don't want anyone leaking this stuff. Not knowing what we are about to discuss . . .oh, well, Bill I guess it is your call. ADVISER #4 should stay. I guess he should . . . huh . . . oh, OK, well, you make the call, Bill.

WM CASEY: OK, I guess ADVISER #4 will stay. But I think [CIA FEMALE TRANSCRIBER] must stay. I'll make that call.

PRESIDENT: OK, I guess we can do our business first, give me a few minutes. Let's get some food first, or snacks. How long will . . . huh . . . oh, about one hour?

The CARETAKER: Mr President, I estimate this will last about one hour, at least the first part of it. This is a very complicated subject to brief. I can do it, but maybe the questions will extend [beyond the first] the time period.

PRESIDENT: OK, I see. Let's take a break and then reconvene.

(Break)

The CARETAKER: OK, Mr President, are we ready?

PRESIDENT: Yes, we are, let's go.

The CARETAKER: Mr President, as was mentioned earlier, I must say, that this briefing has the highest classification within the U.S. government. I will start with a slide presentation. I have most of this briefing on the slides, but I also have an outline that I have passed out to each [person] in attendance.

. . . The United States of America has been visited by extraterrestrial visitors since 1947. We have proof of that. However, we also have some proof that Earth has been visited for many THOUSANDS OF YEARS by various races of extraterrestrial visitors. Mr. President, I'll just refer to those visits as ETs. In July 1947, a remarkable event occurred in New Mexico. During a storm, two ET spacecraft crashed. One crashed southwest of Corona, New Mexico, and one crashed near Datil, New Mexico (see plate 3). The U.S. Army eventually found both sites and recovered all of the debris and one live alien. I'll refer to this live alien as "Ebe1."

PRESIDENT: What does that mean? Do we have codes or a special terminology for this?

The CARETAKER: Mr. President, EBE means "Extraterrestrial Biological Entity." It was a code designated to this creature by the U.S. Army back in those days. This creature was not human and we had to decide on a term for it. So, scientists designated the creature as Ebe1 (see plates 4 and 5). We also referred to it as "Noah." There was different terminology used by various aspects of the U.S. military and intelligence community back then.

PRESIDENT: Do we or did we have others? The number "1" would seem to indicate we had others.

The CARETAKER: Yes, we had others. Back then, the term was EBE and no number designation. We'll explain how the others came into our knowledge.

PRESIDENT: OK, sorry, I was just wondering and I guess, well, I'm sure the briefing will cover this. Please continue.

The CARETAKER: All the debris and EBEs recovered from the first crash site were taken to Roswell Army Air Field, Roswell, New Mexico. EBE was treated for some minor injuries and then taken to Los Alamos National

Laboratories, which was the safest and most secure location in the world. Special accommodations were made for EBE. The debris was eventually transferred to Dayton, Ohio, home of the Air Force Foreign Technology Division. The second crash site wasn't discovered until 1949 by some ranchers. There were no live aliens at this site. All this debris went to Sandia Army Base in Albuquerque, New Mexico.

PRESIDENT: OK, a question, regarding the first site, how many aliens were in the spaceship?

The CARETAKER: Five (5) dead aliens and one (1) alive. The bodies of the dead aliens were transported to Wright Field in Ohio and kept in a form of deep freeze. They were later transported to Los Alamos, where special containers were made to keep the bodies from decaying. There were four (4) dead aliens in the second crash site. Those bodies were in an advanced state of decaying [*sic*]. They had been in the desert for the past two (2) years. Animals and time got to those bodies. The remains were transported to Sandia Base and eventually onto Los Alamos. We determined [that] both crashed spaceships were of similar design and the bodies of the aliens were all identical. They looked exactly the same. They had the same height, weight, and physical features. Here are the photographs of the aliens (Pause) [While presumably the president is looking at the photographs]

PRESIDENT: Can we classify them? I mean can we . . . well, connect them with anything Earthly?

The CARETAKER: No, Mr. President. They don't have any similar characteristics of a human, with exception of there [their] eyes, ears and a mouth. Their internal body organs are different. Their skin is different, their eyes, ears, and even breathing is different. Their blood wasn't red and their brain was entirely different from human. We could not classify any part of the aliens with humans. They had blood and skin, although considerably different than human skin. Their eyes had two different eyelids. Probably because their home planet was very bright.

PRESIDENT: Maybe I'm getting ahead, but do we know where they came from? Mars, our system or where?

The CARETAKER: Yes, Mr. President, we do know where they come from. I can go into this now, or I can wait until it comes up in the briefing.

PRESIDENT: No, no, please, continue. I can wait.

The CARETAKER: Thank you, Mr. President. EBE stayed alive until 1952 when it died. We learned a great deal from EBE. Although EBE did not have voice organs like humans, it was able to communicate with an operation performed by military doctors. EBE was extremely intelligent. It learned English quickly, mainly by listening to the military personnel who were responsible for EBE's safety and care. EBE was housed in a special area at Los Alamos and Sandia Base. Although many different military doctors, scientists, and a select number of civilians studied EBE, it never became upset or angry. EBE helped us learn from all the items found in the two crash sites. EBE showed us how some of the items worked, such as a communications device. It also showed us how various other devices worked.

PRESIDENT: Excuse me, but you are referring to this creature as an IT. Did it have a gender?

The CARETAKER: I'm sorry Mr. President, but yes, it was male. Within EBE's race they had males and females.

PRESIDENT: OK, thank you. Please continue . . .

The CARETAKER: It took the EBE spaceship nine (9) of our months to travel the 40 [38.42] light-years. Now, as you can see, that would mean the EBE spaceship traveled faster than the speed of light. But this is where it gets really technical. Their spaceships can travel through a form of "space tunnels" that gets them from point A to point B faster without having to travel at the speed of light. I cannot fully understand how they travel, but we have many top scientists who can understand their concept.

For those portions of the Reagan briefing in March 1981 relating to Project Serpo, please see appendix 11.

A NEW PLANETARY ERA

The survival of Ebe1, what we learned from him and from the crashed discs, and our subsequent association with his planet opened up an entirely new era of our planet's history. It was the first baby step for Earth on the stage of galactic affairs, and it initiated a lasting alliance between the United States and a civilization on a distant planet. This was our antidote to the threat of the Fourth Reich in Antarctica, and it came at a perfect time. Roswell transformed American technology, expanded our cosmic vision, and opened the door to the space age.

4

LOS ALAMOS

It was sometime afterward when the thought flashed upon my mind that the disturbances I had observed might be due to an intelligent control. Although I could not decipher their meaning, it was impossible for me to think of them as having been entirely accidental. The feeling is constantly growing on me that I had been the first to hear the greeting of one planet to another.

Nikola Tesla, "Talking with the Planets," *Colliers Weekly* (March 1901)

The selection of the Los Alamos National Laboratory near Santa Fe, New Mexico, as the site to house the alien found alive at the Corona crash site, seemed at first glance to be rather strange and inappropriate. At the time of the Roswell crash, in July 1947, less than two years after the two atomic bombs had been dropped on Japan, and only four years after it was founded, the Los Alamos Laboratory was still somewhat primitive. Originally the Los Alamos Ranch School, a private school for boys who wanted to live an outdoor life, it was selected by J. Robert Oppenheimer, the director of the Manhattan Project, and approved by General Leslie Groves, and then was commandeered by the Army in November 1942 for the express purpose of designing and developing the atomic bomb. That is to say, it was officially condemned by the government so that the property could be acquired pursuant to a military purpose. In order to allow the boys to finish the fall semester, Secretary of War Henry Stimson agreed to wait until February 8, 1943, to take possession.

During the war, it was staffed primarily by high-level theoretical physicists from several different countries. Immediately after the war, in the fall of 1945, the scientific ranks were severely reduced as the major nuclear scientists returned to academia and corporate consulting, and the lower-level workers left to pursue advanced degrees. Oppenheimer resigned a few weeks after the war ended to become director of the Institute for Advanced Study at Princeton, and Norris Bradbury succeeded him as the director in October 1945. By spring 1946, only twelve hundred staff members remained, as Bradbury sought to define the new civilian role of the laboratory in the postwar world, now working closely with the nascent Atomic Energy Commission. But the exclusive focus of the laboratory in the new era continued to be the development of atomic weaponry. According to the Los Alamos website:

> By the late 1940s, funding was secured to rebuild the main technical area and improve housing, as the Laboratory refined and tested fission weapons and gradually expanded the hydrogen bomb research program. Two test series, Operations CROSSROADS and SANDSTONE, were conducted in 1946 and 1948, respectively. Six nuclear weapons tests were conducted in the 1940s, allowing the nation's stockpile to grow from two bombs, in late 1945, to 170 in 1949.

Interestingly, Major Jesse Marcel, the base security officer at Roswell, had been the Army security officer for Operation CROSSROADS conducted at the Bikini Atoll. There is no evidence that any other type of R&D was conducted at the laboratory in that period. Given the limited resources available to him during the late 1940s, it seems highly improbable that Bradbury was financially able to pursue any other type of research.

Los Alamos National Laboratories, present day

PARANOIA RULES

It seems that the most important rationale for sending the sole alien survivor of the Roswell crash to Los Alamos in 1947 must have been based on trying to learn whatever they could about the advanced technology incorporated into the alien craft, and whatever other scientific information he could give them. Certainly, the theoretical physicists at the Laboratory would have been most capable of comprehending that information and possibly converting it to useful human technology. In a more perfect world, an unexpected visitor from another planet might have been sent to a top university where Earth academicians would have attempted to learn about his home world. But, only two years after a brutal, devastating war, and with the military expecting another one, advanced technology adaptable to weaponry was the main preoccupation of the U.S. government. Paranoia superseded civilized curiosity.

But from another viewpoint, that paranoia may have been fully justified, since we knew nothing about the aliens' motives. After all,

*Cielo supercomputer at Los Alamos**

this craft was evidently scrutinizing the most sensitive military installation in the world. If an alien civilization was contemplating an invasion of Earth, it would certainly begin by checking out the most powerful military capability of the most powerful nation on the planet. It is supremely logical that the military would have come to that conclusion, and that possibility alone justified sending the alien to Los Alamos. In fact, as more and more of the secret information about the Roswell crash emerged over time, it began to appear that the alien craft may indeed have been conducting a spy mission in advance of some sort of mass landing. Army Colonel Philip Corso apparently believed that when he wrote in his book, *The Day after Roswell:*

> The real truth behind a fifty-year history of a war that looked like the ultimate defeat of humankind . . . can now finally be told because we prevailed. It was because in the dark hours before dawn in July 1947 the army, only dimly recognizing that we were on the edge of a potential cataclysmic event, pulled the crashed spacecraft

*Los Alamos has always had one of the most powerful computer systems in the world. Its newest was installed in 2011 and is used jointly by Lawrence Livermore Laboratories and Sandia. The Cielo, built by Cray, can process more than a quadrillion floating point operations per second! It consists of 96 cabinets and 300 terabytes of memory.

> out of the desert and harvested its parts just like the inhabitants of that vehicle wanted to harvest us.

This dramatic supposition was evidently based on a top secret report that human body parts had been found on the crashed alien craft!

And then, of course, there were the security considerations. Since the government had immediately decided, right after the crash, to keep the entire matter in top secret containment, the site chosen to host the alien had to be extremely secure, and no facility was more secure, at that time, than the home of the Manhattan Project. That requirement alone ruled out a university location. Even the Pentagon itself, which was only four years old in 1947, did not have as high a level of security as Los Alamos in those early years, since the Pentagon was right in the middle of a very busy section of northern Virginia. By contrast, the only entrance to Los Alamos was over a single mountain road tucked inside a canyon.

THE FIRST MESSAGES

The major concern in dealing with the alien was, first and foremost, communication. And evidently it was believed that if the most brilliant scientists in the country could not learn to communicate with the ET, then nobody could. Furthermore, once the alien was in a secure location, other, more appropriate language personnel could be brought in, as needed, to facilitate communication and interaction. And they, of course, would be thoroughly investigated and made to sign security oaths before gaining admission.

MJ-12 decided to refer to the Roswell aliens as "Ebens." This was simply an unimaginative derivation of EBE, standing for "extraterrestrial biological entity." After being comfortably ensconced at Los Alamos, the alien, who was given the designation Ebe1 by MJ-12, willingly cooperated in attempting to overcome the communication barrier. But the language differences were huge and seemed insurmountable. As shown in *Close Encounters of the Third Kind,* Steven Spielberg's 1977 movie (see chapter 18), the Eben language consisted of tonal variations, and sounded almost musical. One contributor to the Serpo website described it as "high-pitched singing." Some of the sounds were not even reproducible by the Americans. For the entire five-year period that Ebel remained alive, he was only able to teach the Los Alamos scientists about 30 percent of the Eben language. Anonymous reported that "Complex sentences and numbers could not be recognized."

Ebel identified a piece of equipment found intact on the alien craft as a communications device for sending and receiving messages from his home planet, and he showed the scientists how to use it, but they couldn't get it to work. So, no communications with his planet were possible for five years, until just prior to his death in the summer of 1952, when one of the scientists realized that the device had to be powered by the energy source on the alien craft. When they tried that, it worked. What is very surprising about this is the fact that it was an earthling who figured that out, not the alien. Evidently, Ebel just wasn't very smart. Anonymous tells us that Ebel was a mechanic. He wasn't a scientist. Once the connection was made, the alien commenced sending messages to his planet. During the summer of 1952, he sent six messages, all of which were successfully transmitted. He was able to roughly translate the messages into English for the benefit of the Los Alamos scientists. Message #1 notified his planet that he was still alive. Message #2 told about the crash and the fact that all of his crew were killed upon impact. The third message requested that a rescue craft be sent to Earth to pick him up. At the prompting of the scientists, the fourth message suggested that a formal meeting be arranged with Earth officials, who would, of course, be American. And then we are told by Anonymous that message 5 said that the United States government had requested an exchange program. Finally, the sixth message gave the landing coordinates on Earth for any future visitation. Why this was attempted is puzzling, since Ebel really couldn't interpret our chronology or numbering system, and we already knew that those coordinates were probably not going to be understood by the Ebens anyway.

Ebel did receive some replies to his messages, but they were replies that only he understood. His attempted translations were confusing. Apparently, his planet agreed to a return visit, but the date they specified was over ten years away! Our people concluded that this must have been a mistake, but they were unable to obtain a clarification before Ebel died in the late summer of 1952.

In a parenthetical note from Anonymous about the exchange suggestion in message 5, he says:

> It is believed, but NOT documented, that Ebe1's U.S. military caretaker had suggested to Ebe1 that an exchange program be set up which would allow our people to visit and exchange culture, scientific information, and collect astronomical [data] during a space trip by an American military team or what eventually became known as the team members.

As noted above, Ebel did indeed make that suggestion, but apparently there was no reply to that message. That proposal for an exchange program this early in the game was evidently purely in the interests of intelligence, and not goodwill, since the apparent spy mission of the aliens, and the discovery of human body parts on the spacecraft very likely caused us to view the Ebens as potential invaders, even though we already knew that they were not human flesh eaters! Nevertheless, that discovery probably created sufficient distrust for us to want to obtain inside information about this race that dropped onto our planet out of the skies. After all, they were caught stealthily gathering information about our military capability instead of landing at the United Nations in Manhattan and asking to be taken to our leader. Furthermore, they had the technology to reach our planet, and be assured, we were going to do whatever it took to get our hands on that technology. So, if we were able to walk around on their planet, and they were willing to let us visit, we were going to go there! The fact that the suggestion originally came from the "military caretaker" strongly implies that the proposal was based on a military consideration.

WHAT HATH LOS ALAMOS WROUGHT?

After the death of Ebel, the Los Alamos scientists continued to try to establish communication with the Ebens' planet. They had a lexicon of the Eben written language to work from, provided by Ebe1. According to the DIA information, the scientists sent several messages in 1953 that went unanswered. And then, after an intense, eighteen-month effort to improve their syntax, they sent two messages in 1955, and finally received a reply. This was an amazing breakthrough for our planet, and an incredible accomplishment by the scientists. We had begun an actual dialogue with an alien civilization across a vast ocean of space. Far more significant than the accomplishments of Guglielmo Marconi and Alexander Graham Bell, if this event had not been so secret, it would have generated huge headlines in all the major newspapers of the world. One can imagine the celebratory scene at the Laboratory when that first alien message appeared on the communication device. Now came the job of translation. Unaided, the scientists could only comprehend about 30 percent of the message. However, with the help of language special-

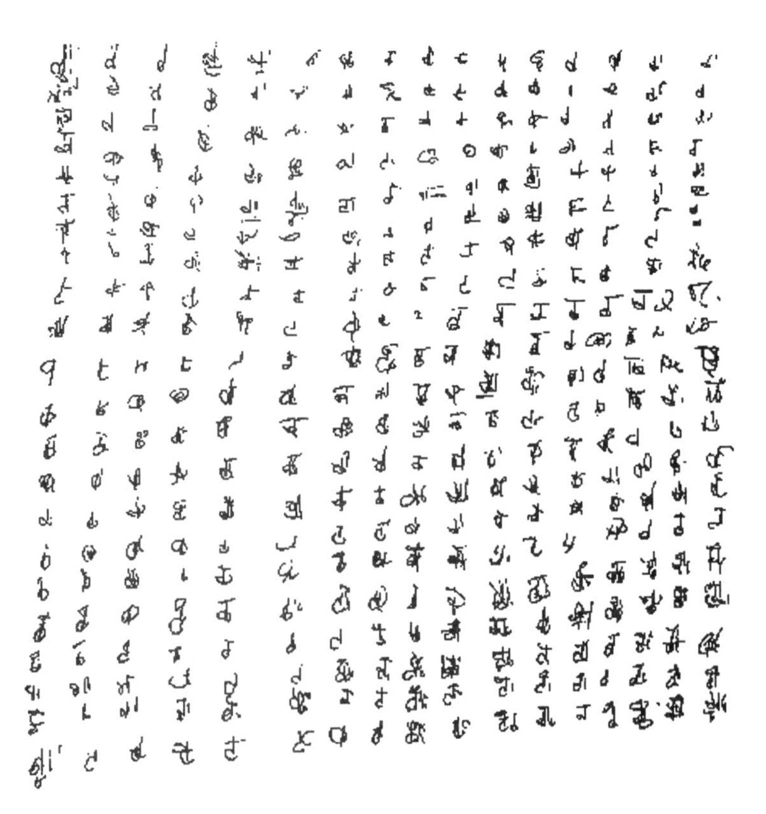

The Eben alphabet

ists from both U.S. and foreign universities, they were able to translate most of the message.

Based on the probability that the Ebens are smarter than we are, the scientists then decided to send a reply in English, hoping the Ebens would find it easier to translate our language. They received a reply in broken English about four months later. The aliens didn't understand the concept of verbs, so their message contained only nouns and adjectives. It took several months for us to figure out the English reply. It became clear that if we sent them some basic English lessons, it might be possible to carry on a productive dialogue much faster than if we had to continue to labor over their impossibly cryptic language. This was done, and six months later Los Alamos received another English message that was much more comprehensible. They were catching on, but, according to Anonymous, "Ebens were confusing several different English words and still failed to complete a proper sentence." But it was an auspicious beginning. The Ebens now had the basics for English communication. Certainly, if they had the ability to travel throughout the galaxy and interact with other civilizations, they could decipher the rules of a language that human fifth-graders mastered routinely. Very sensibly, the Ebens then sent us a compendium of their alphabet with what they believed to be corresponding English letters. This was turned over to the university linguists working with the Los Alamos scientists. Our language specialists struggled with it and had a very difficult time. It took another five years before we acquired a basic understanding of the Eben language, and the Ebens became somewhat, haltingly able to communicate in English.

A RETURN VISIT

During that five-year period, which constituted the final years of the Eisenhower administration, the scientists continued to seek to arrange a return visit of the Ebens to Earth. Apparently, the desire for another visit was mutual. It appeared that we both wanted to establish some sort of diplomatic relationship. As previously mentioned, while we—especially our scientists—were no doubt motivated by high-minded sociological and scientific interests, our government officials and military and intelligence people were suspicious of the aliens' intentions, and were more concerned with understanding their advanced technology, particularly as it applied to weaponry. We were probably still hopeful that a return visit would lead to an exchange program. It will be recalled that this was proposed by Ebe1 in his fifth message, at our prompting, but if he did receive a reply agreeing to this, he wasn't able to translate it for us. Very probably, the aliens had the same goals. We can safely conclude that they were very much interested in our atomic bombs. As we later learned while on their planet, they had not developed atomic energy, although they did have the more powerful particle-beam weaponry, and had used it in war.

Furthermore, the Ebens wanted to retrieve the bodies of their dead comrades. This involved some complications. While we did keep the remains of the bodies frozen at Los Alamos, using some very advanced cryonic technology, Anonymous says that we performed autopsies on four of the five dead aliens found at the Corona crash site. This was probably explained to them, but should not have surprised or shocked them at all. In fact, they probably expected it, and, as we later learned on Serpo, the Ebens had a rather ghoulish biotechnology research operation of their own that went far beyond simple autopsies.

Planning the return visit turned out to be much more complicated than we had expected. We were unable to understand their date and time system, and they could not comprehend ours. We sent them all the details of our planetary rotation and revolution, how we marked dates and time, and precisely where we were at the moment we sent the data. But the Ebens were never able to understand our system. Finally, by 1960, we figured out theirs, and we were able to send them longitude and latitude coordinates that we thought they could understand; however, we couldn't be sure. In early 1962 we developed a better arrangement. Anonymous says, "We then decided to just send pictures showing the Earth, landmarks, and a simple numbering system for time periods." They allowed us to select a date, and we chose April 24, 1964. Selecting a landing location proved to be even more complex. The military planners wanted to make absolutely certain that security would not be compromised, and that the press or the public would not even get a hint of what was going on. At first they considered remote islands, but then realized that unusual movements of naval vessels could arouse suspicion. They decided that the site had to be controlled by the military to ensure complete secrecy. Finally, they settled on the southern end of the White Sands Missile Range, near Holloman Air Force Base in New Mexico. Holloman was previously known as Alamogordo Army Air Field, a training site for Eighth Air Force bombardment crews during World War II. They also selected a fake location on the base itself to misdirect interest. These decisions were made, and confirmed by the Ebens, in March 1962. It had taken ten years from the time of the death of Ebe1 to arrive at this historic agreement.

AN INCOMPLETE STORY

This long-range planning schedule is somewhat surprising, since the Ebens must have had a mother ship already in orbit around Earth, and consequently must have been able to send scout ships to the surface without delay. Certainly, that tiny six-man craft that crashed near Roswell did not travel to Earth by itself over thirty-eight light-years from the Zeta Reticuli system, which we later determined to be their home constellation. And if a mother ship had remained in orbit around our planet after the Roswell crash, why was it not possible for the Ebens to just send another scout craft to the surface to arrange the formal rendezvous directly? Certainly, MJ-12 and the government must have wondered about that and must have asked that question over the communication device at an early stage in the interplanetary dialogue.

This brings up the possibility that this did indeed happen—that some negotiations may have ensued without the subsequent knowledge of the DIA, the agency that ultimately released the Serpo material in 2005. This should come as no surprise, since the DIA did not exist in 1953. The Defense Intelligence Agency was created by President John F. Kennedy's secretary of defense, Robert S. McNamara, in October 1961. Consequently, whatever happened between the Roswell crash in July 1947 and the advent of the DIA in 1961 could only have been known to them from information they were able to gather after their creation, from top secret Army and Air Force intelligence sources operating under President Eisenhower. And it would be fair to conclude that these military intelligence agencies were not very happy about having to turn this information over to a brand-new intelligence agency created by a new, young, Democratic president, who planned to replace them by combining all their functions under a single umbrella organization. Consequently, they were probably not very cooperative, and may have provided incomplete information, or possibly even disinformation, about the era when Eisenhower was president. Most likely, they convinced MJ-12 itself not to share that intelligence with the DIA in the interests of compartmentalization. As will be seen in chapter 5, this is probably exactly what happened, since we now know that the Ebens did send another scout craft to Earth on May 20, 1953. And that craft did not crash—it landed. Evidently, those messages that were sent in early 1953 must have been answered after all, but the DIA was not informed about this when the agency joined the project in 1962.

5

KINGMAN

Just as with the Roswell story, the revelation that an alien craft was thought to have crashed near Kingman, Arizona, in 1953 was kept tightly contained for almost twenty-five years. And, if it hadn't been for the very thorough reporting of a dedicated NICAP (National Investigations Committee on Aerial Phenomena) investigator, it probably would have remained buried for at least another ten years. The dogged and indomitable digging of the NICAP UFO teams in the 1950s and 1960s, and their rigid screening and reporting protocols, have become legendary. Time and again, their facts were proven incontrovertible, and their reports were unassailable. And it was this professionalism that most impressed the newspapers, and convinced them to give the UFO story mainstream coverage in the 1960s. As a result of that publicity, the government was embarrassed into forming its own formal public UFO investigative organization in 1966—the now-infamous Condon Committee, which relied heavily on the fieldwork and reports of NICAP. Author and veteran NICAP investigator Raymond E. Fowler first broke the Kingman story in an article in *UFO Magazine* in 1976. This article was then incorporated into his book, *Casebook of a UFO Investigator: A Personal Memoir,* published by Prentice-Hall in March 1981. J. Allen Hynek, the chief scientific consultant for the Air Force Project Bluebook, said about Fowler "an outstanding UFO investigator . . . I know of no one who is more dedicated, trustworthy or persevering. . . . [Fowler's] meticulous and detailed investigations . . . far exceed the investigations of Bluebook" (www.crowdedskies.com/ray_fowler_bio.htm).

UFO investigator Raymond E. Fowler

In his article (and in his book), Fowler reports that, in the course of his NICAP investigations in 1973, he interviewed someone regarding a crashed disc found in Arizona. The interviewee chose to remain anonymous in Fowler's report, but agreed to go by the assumed name Fritz Werner. Always thorough in obtaining evidence, Fowler asked Werner for a signed statement giving all the details of his experience, which he agreed to. In that statement, dated June 7, 1973, Werner testified that on May 21, 1953, he "assisted in the investigation of a crashed unknown object in the vicinity of Kingman, Arizona" (see plate 6). In the statement, Werner described the object as oval, and about thirty feet in diameter. He said it was constructed of "an unfamiliar metal which resembled brushed aluminum." He said further that the disc was undamaged, and had penetrated only about twenty inches into the sand. This would suggest that it came down at a very moderate speed, implying that perhaps this really wasn't a crash at all, but more of a landing. A hatch about 3½ feet high was open, another indication of an orderly landing and egress. The hatch height suggested that the occupants of the craft were about that tall. As part of his statement, Werner claimed that he saw a dead alien in a tent near the craft. In his own words, as reported by Fowler, he said, "An armed military policeman guarded a tent pitched nearby. I managed to glance inside at one point, and saw the dead body of a four-foot human-like creature in a silver metallic-looking suit. The skin on its face was dark brown." His testimony becomes slightly questionable at this point because, in his description, he stated with confidence that this alien had been the "only occupant." Since he was not permitted to enter the craft, there was no way he could have known by observation how many occupants had been inside. Consequently, somebody in authority must have told him that. This "slip" opens the entire statement up to the possibility of disinformation. It suggests that he was deliberately allowed to notice the dead alien, and then was told that it was the only occupant.

Werner told Fowler that he was one of about fifteen engineers and scientists brought to the crash site in a bus with blackened windows from Phoenix Sky Harbor Airport, on the night of May 21, 1953. He had been working on loan to the Atomic Energy Commission at the Nevada Test Site to assess the damage to various structures due to atomic test blasts. He was flown into Sky Harbor from the Indian Springs, Nevada, Air Force Base. At the Kingman site, given his particular expertise, he was expected to estimate the velocity of the disc at impact from the way it was embedded in the soil. He ultimately concluded that it had been traveling at about 100 knots (115 mph). That could almost be considered landing speed for an antigravity craft. Fowler knew Werner's real name, which we now know is Arthur G. Stancil, and did check him out meticulously. He found that in 1953 Stancil was indeed a project engineer with the Air Force under contract to the Atomic Energy Commission. He had previously worked at Wright-Patterson Air Force Base in the Foreign Technology Division. He determined that Stancil's history and all his credentials were genuine, and that he had no reason to try to perpetrate a hoax.

Each of the people on the bus was escorted one at a time, by military policemen from the bus to the heavily guarded, brightly illuminated site, and instructed to just focus on his or her specific job. And they were told not to engage in any fraternization or discussion about what they learned. So really Werner was violating the instructions when he peered inside the tent. Given the rigid security measures governing the entire procedure, it is highly unlikely that the MPs failed to prevent Werner from looking into the tent. More likely, he was allowed to see the alien, and to go away believing that he had stolen a furtive glance, and was then led to believe that he had seen the only occupant, and that occupant was dead. In fact, it seems unlikely that this staged scene was only for Werner's benefit. Very probably, all fifteen investigators had the same opportunity.

And when it came to secrecy vows, all the participants received very lightweight treatment. An Air Force colonel got on the bus and had them all sign the Official Secrets Act. He then asked them all to raise their right hand and take an oath to never reveal what they had experienced. This was very different from the horror stories others have talked about, where terrifying intimidation and death threats were common. It seems likely that the Air Force expected and wanted these people to later reveal the event and what they had seen in the tent so that the "sole dead occupant" version would gain currency. We will see below why this tricky bit of manipulation might have been carried out, and why it was so important.

A DIFFERENT VERSION

The website that hosted the original Project Serpo revelation, www.serpo.org, invited military, intelligence, and other government insiders who had direct knowledge about the Serpo story to send in what they knew, or had experienced, for publication on the website. One such contribution published on the site in August 2006 sheds new light on what actually happened at Kingman. It is a photographic image of a two-page classified memo, evidently written as a briefing document for another government group. It is dated March 24, 1995. It's interesting to note here that whoever sent it in evidently understood, or perhaps suspected, that there was a connection between the Kingman event and Project Serpo. It was sent by a researcher who supplied his name, but remained unidentified in the posting. In the typed copy, the document makes reference to a "vessel that fell in Arizona." Written in longhand, at the top it says, "1953 Kingman Cerbat Mountains north of the Kingman Army Air Base." The Army air base is now the Kingman Municipal Airport. The Cerbat Mountains are about ten miles northwest of the airport.

The document gives fascinating insider details about the Kingman crash retrieval as part of an explanation of how the long-term relationship with the Ebens developed, and how the reverse-engineering program began. To begin with, it is a confirmation of Stancil's report that an alien craft came down near Kingman in 1953. The major departure

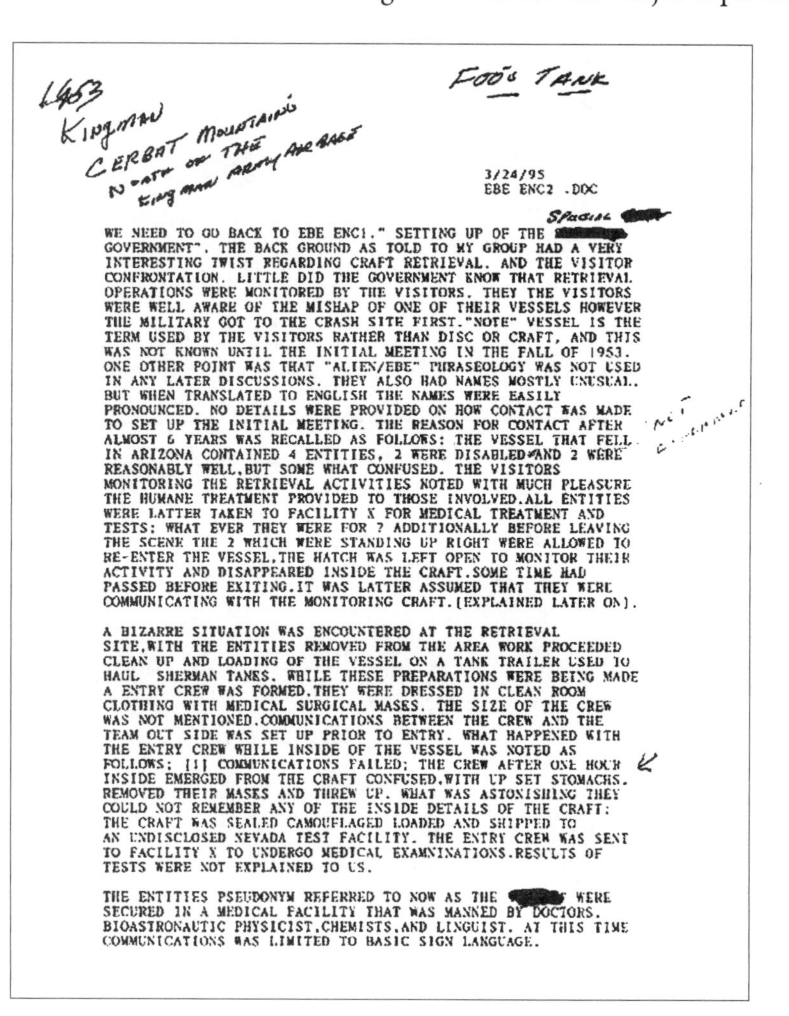

3/24/95
EBE ENC2 .DOC

WE NEED TO GO BACK TO EBE ENC1." SETTING UP OF THE [redacted] GOVERNMENT". THE BACK GROUND AS TOLD TO MY GROUP HAD A VERY INTERESTING TWIST REGARDING CRAFT RETRIEVAL. AND THE VISITOR CONFRONTATION. LITTLE DID THE GOVERNMENT KNOW THAT RETRIEVAL OPERATIONS WERE MONITORED BY THE VISITORS. THEY THE VISITORS WERE WELL AWARE OF THE MISHAP OF ONE OF THEIR VESSELS HOWEVER THE MILITARY GOT TO THE CRASH SITE FIRST."NOTE" VESSEL IS THE TERM USED BY THE VISITORS RATHER THAN DISC OR CRAFT, AND THIS WAS NOT KNOWN UNTIL THE INITIAL MEETING IN THE FALL OF 1953. ONE OTHER POINT WAS THAT "ALIEN/EBE" PHRASEOLOGY WAS NOT USED IN ANY LATER DISCUSSIONS. THEY ALSO HAD NAMES MOSTLY UNUSUAL. BUT WHEN TRANSLATED TO ENGLISH THE NAMES WERE EASILY PRONOUNCED. NO DETAILS WERE PROVIDED ON HOW CONTACT WAS MADE TO SET UP THE INITIAL MEETING. THE REASON FOR CONTACT AFTER ALMOST 6 YEARS WAS RECALLED AS FOLLOWS: THE VESSEL THAT FELL IN ARIZONA CONTAINED 4 ENTITIES, 2 WERE DISABLED AND 2 WERE REASONABLY WELL,BUT SOME WHAT CONFUSED. THE VISITORS MONITORING THE RETRIEVAL ACTIVITIES NOTED WITH MUCH PLEASURE THE HUMANE TREATMENT PROVIDED TO THOSE INVOLVED.ALL ENTITIES WERE LATTER TAKEN TO FACILITY X FOR MEDICAL TREATMENT AND TESTS; WHAT EVER THEY WERE FOR ? ADDITIONALLY BEFORE LEAVING THE SCENE THE 2 WHICH WERE STANDING UP RIGHT WERE ALLOWED TO RE-ENTER THE VESSEL,THE HATCH WAS LEFT OPEN TO MONITOR THEIR ACTIVITY AND DISAPPEARED INSIDE THE CRAFT.SOME TIME HAD PASSED BEFORE EXITING.IT WAS LATTER ASSUMED THAT THEY WERE COMMUNICATING WITH THE MONITORING CRAFT.(EXPLAINED LATER ON).

A BIZARRE SITUATION WAS ENCOUNTERED AT THE RETRIEVAL SITE,WITH THE ENTITIES REMOVED FROM THE AREA WORK PROCEEDED CLEAN UP AND LOADING OF THE VESSEL ON A TANK TRAILER USED TO HAUL SHERMAN TANKS. WHILE THESE PREPARATIONS WERE BEING MADE A ENTRY CREW WAS FORMED.THEY WERE DRESSED IN CLEAN ROOM CLOTHING WITH MEDICAL SURGICAL MASKS. THE SIZE OF THE CREW WAS NOT MENTIONED.COMMUNICATIONS BETWEEN THE CREW AND THE TEAM OUT SIDE WAS SET UP PRIOR TO ENTRY. WHAT HAPPENED WITH THE ENTRY CREW WHILE INSIDE OF THE VESSEL WAS NOTED AS FOLLOWS; (1) COMMUNICATIONS FAILED; THE CREW AFTER ONE HOUR INSIDE EMERGED FROM THE CRAFT CONFUSED,WITH UP SET STOMACHS. REMOVED THEIR MASKS AND THREW UP. WHAT WAS ASTONISHING THEY COULD NOT REMEMBER ANY OF THE INSIDE DETAILS OF THE CRAFT: THE CRAFT WAS SEALED CAMOUFLAGED LOADED AND SHIPPED TO AN UNDISCLOSED NEVADA TEST FACILITY. THE ENTRY CREW WAS SENT TO FACILITY X TO UNDERGO MEDICAL EXAMNINATIONS.RESULTS OF TESTS WERE NOT EXPLAINED TO US.

THE ENTITIES PSEUDONYM REFERRED TO NOW AS THE [redacted] WERE SECURED IN A MEDICAL FACILITY THAT WAS MANNED BY DOCTORS. BIOASTRONAUTIC PHYSICIST,CHEMISTS,AND LINGUIST. AT THIS TIME COMMUNICATIONS WAS LIMITED TO BASIC SIGN LANGUAGE.

Memo sent to the Project Serpo website

Cerbat Mountains near Kingman, Arizona

from his report concerns the number of occupants and their condition. The document says that there were four, all alive, claiming "The vessel . . . contained 4 entities, 2 were disabled and 2 were reasonably well, but some what [sic] confused." The writer of this document claims to know that another alien craft was actively monitoring the craft retrieval, although our people were not aware of that at the time. He says, "Little did the government know that the retrieval operations were monitored by the visitors . . . the visitors were well aware of the mishap of one of their vessels, however the military got to the crash site first." The choice of the word "mishap" is very interesting in this context. It infers that the problem was not so much that the craft crashed, but that the landing place was not the intended location. It implies that the craft missed its destination. Considering that the retrieved disc was ultimately taken to the Nevada Test Site, it is reasonable to conclude that it was originally heading for that location, only about two hundred miles away on a direct northwest flight path.

The likelihood that this rendezvous was prearranged explains why the military retrieval crew got there so quickly. The over-the-road military response had to be lightning-fast in order to be ahead of the supersonic alien craft. Evidently, the retrieval team had been on alert throughout the Arizona-Nevada area, anticipating that just such a "mishap" might occur. What is most amazing about this rapid response is the fact that it could only have been the result of some sort of communication. This was not a case of some civilian report to the police, who then contacted the military. That would have taken hours, perhaps days. It had to be a direct message giving the geographic coordinates of the crash, since it occurred in a remote, mountainous area. It implies that the military had been in direct communication with the aliens. Since we already know that the Los Alamos scientists had a direct link to the alien planet, the sequence of events must have involved a report back to Serpo by the stranded crew, followed by a message to Los Alamos from Serpo, followed by an immediate message from Los Alamos to the response team.

According to the memo, it was later determined that the aliens in the monitoring craft had been very pleased with the humane treatment that we provided to the grounded disc crew. That means that obviously we had cordial relations with these ETs after the landing. That's the only way we could have received that feedback. The memo says that the aliens "had their own agenda. They were busy doing there [*sic*] own analysis of items provided in the quarantined area set up. This included: food, facilities, and us. It begin [*sic*] to appear that we were the captees [*sic*] and they were the captors." The two uninjured aliens requested that they be allowed to reenter the craft. We agreed, but left the hatch open so that they could be observed. It was later realized that they were probably communicating with the monitoring craft. After they exited, all four aliens were taken away to a special habitat that we had already prepared for them in Los Alamos Laboratories, where they were given medical treatment and tests. The memo says that they "were secured in a medical facility that was manned by doctors, bioastronautic physicist[s] chemists, and linguist[s]. At this time communications was limited to basic sign language." Again, that high degree of preparation clearly demonstrates that the entire operation was prearranged. That habitat at Los

Alamos was evidently the same one created to house Ebel before his death one year earlier. The fact that the Kingman aliens were brought to the very same facility clearly implies that they were also Ebens.

A ROCKET SHIP TO HELL

In the memo, we learn of an interesting incident that occurred after the aliens had been taken away to Los Alamos. The military retrieval team decided to enter the craft. What followed was unexpected and bizarre. According to the memo, "a[n] entry crew was formed, they were dressed in clean room clothing with medical surgical masks. The size of the crew was not mentioned. Communications between the crew and the team out side [*sic*] was set up prior to entry. What happened with the entry crew while inside of the vessel was noted as follows: communications failed; the crew after one hour inside emerged from the craft confused, with up set [*sic*] stomachs, removed their masks and threw up. What was astonishing [was] they could not remember any of the inside details of the craft. When six months later, a member of that original entry crew, a fighter pilot, was asked if he would like to be part of a new entry crew, he replied, 'I would rather take a rocket ship to hell than to go back into that craft.'" That experience clearly connects the Kingman craft with the alien ship that took the astronauts to Serpo. Several members of that team had the identical reaction when they first entered the Eben craft.

What is striking about this retrieval operation, according to the memo, is the fact that the military team came so completely equipped. They had with them a tank trailer used to transport Sherman tanks. That means that they had to know in advance the approximate size of the alien craft, which was exactly thirty feet in diameter. So obviously, the purpose of the entire incident was to give us the disc. Evidently, the aliens were on their way to the Nevada Test Site to land it there and then be taken to Los Alamos by bus. When they landed prematurely near Kingman, we sent the tank trailer. The crane on the tank retriever was able to lift the disc easily and load it onto the trailer. However, they were concerned about the overhang, which might have made the load too wide for city streets. But when they tried to tilt the craft, it proved impossible. So, according to the memo, "Finally it was decided to use the house method of movement utilizing roadblocks secured by military vehicles." This was done, and the alien craft was taken to the Nevada Test Site.

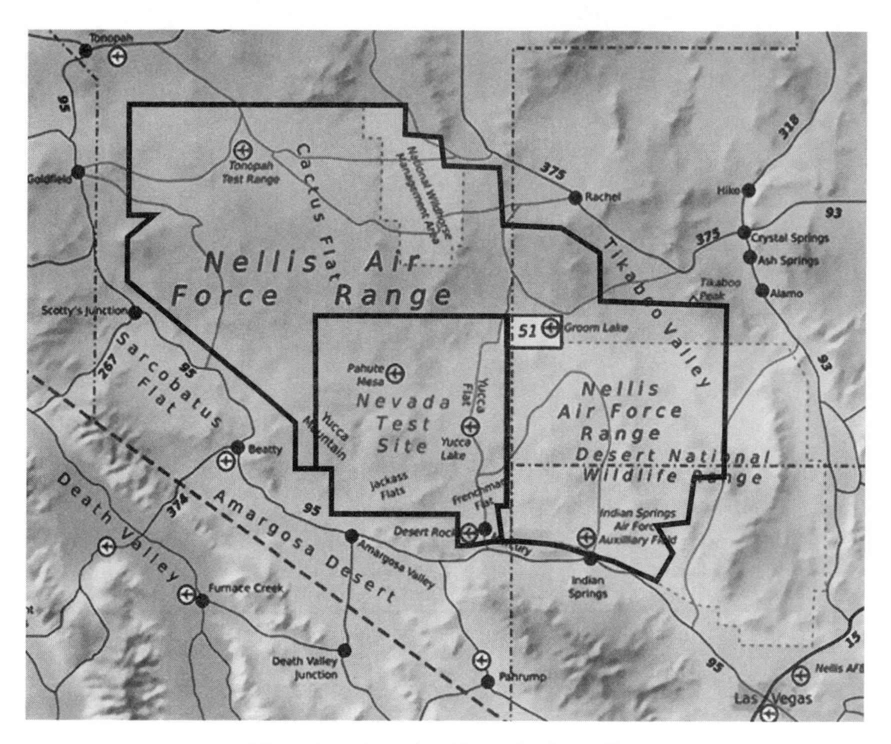

Map showing the Nevada Test Site; note its proximity to Area 51

The scientists at the test site were frustrated by the reentry problem. Also, they found it difficult to concentrate because of a low humming sound that continued to emanate from inside the disc, and since they couldn't enter, they couldn't find the cause. Consequently, for six months they made no progress in analyzing the craft. So, since the four Ebens at Los Alamos wanted to return to the craft, it was decided to bring them to the Nevada Test Site. By this time, the tallest alien and the bioastronautic engineer had established a rudimentary form of communication. The four aliens entered the craft and after a few minutes the hum was silenced. Upon emerging, the tall alien asked the bioastronautic engineer to accompany him back inside. This was permitted. The memo says, "After sometime [*sic*] passed, both made their exit. The engineer looked well and smiling." After that event, the Ebens' preferences were honored. They were allowed to be housed at the test site near their craft, and their requests for additional material, equipment, and literature were granted. Thus began a new phase of human-alien cooperation that allowed the reverse-engineering program to begin in earnest. That would have been sometime around November 1953.

In attempting to reconstruct the sequence of events that occurred after the alien craft landed near Kingman, it becomes clear by putting the two versions together that the scene that was staged for the benefit of the investigators must have taken place after the four live aliens had been packed off to Los Alamos. The dead Eben seen by Stancil could easily have been one of the ten that had been preserved cryogenically at Los Alamos, and was brought to the site on dry ice just to participate in the tableau to be observed by the fifteen investigators—a nifty piece of stagecraft presented by an ad hoc U.S. Air Force theater group!

See appendix 10 for a personal statement by Bill Uhouse, an American engineer who worked on duplicating the alien disc recovered at the Kingman site and building a disc simulator for training our pilots.

6

KENNEDY

Many years ago, the great British explorer George Mallory, who was to die on Mount Everest, was asked why did he want to climb it. He said, "Because it is there." Well, space is there, and we're going to climb it, and the moon and the planets are there, and new hopes for knowledge and peace are there. And, therefore, as we set sail we ask God's blessing on the most hazardous and dangerous and greatest adventure on which man has ever embarked.

John F. Kennedy, speech at Rice University,
September 12, 1962

President Kennedy was catapulted into the middle of world-shaking events in his very first year in office. Perhaps most important was Vostok 1, the successful orbital flight of Soviet cosmonaut Yuri Gagarin on April 12, 1961 (see plate 7). Even though Gagarin was in space for a short 108 minutes, and made only a single orbit, this was a shot across the bow for the United States because we weren't even close to matching that accomplishment, even though we had Wernher von Braun on our NASA team. Kennedy was galvanized by that event. Eight days later, he shot off a memo to Vice President Lyndon B. Johnson, whom he had appointed as chairman of the Space Council, that asked the question, "Do we have a chance of beating the Soviets by putting a laboratory in space, or by a trip around the moon, or by a rocket to land on the moon, or by a rocket to go to the moon and back with a man? Is there any other space program which promises dramatic results in which we could win?" Kennedy was deeply committed to his New Frontier initiative, and space was the new frontier. Space travel was tops on his agenda, and he wasn't going to settle for second best to the Soviets. It clearly demonstrates the importance of this subject to Kennedy when it is realized that this memo to Johnson was sent only three days after the failed Bay of Pigs invasion of Cuba. One would think that explosive subject would have been uppermost in his mind. Just the previous day, several members of the invasion force had been executed by the Castro regime.

We now know from several sources, including Jacqueline Kennedy, how severely shaken Kennedy was over the Bay of Pigs debacle. And yet, only one month later, on May 25, 1961, he delivered his famous man-on-the-moon speech to a Joint Session of Congress, demonstrating, once more, his strong commitment to American triumphs in space. Kennedy's confidence that we could put a man on the moon by the end of the decade was based on von Braun's analysis, which had been solicited by Vice President Johnson. In von Braun's April 29 response to Vice President Johnson's inquiry he said, "We have an excellent chance of beating the Soviets to the first landing of a crew on the moon (including return capability, of course) . . . With an all-out crash program I think we could accomplish this objective in 1967/68."

Metaphoric illustration of the Kennedy fallout over the Bay of Pigs debacle

President Kennedy and Wernher von Braun at the Marshall Space Flight Center at Redstone Arsenal, Huntsville, Alabama, 1963

In his memo, von Braun also discussed funding the development of a nuclear rocket as a long-term goal for going beyond the moon to the exploration of space. In his speech to Congress, Kennedy asked for approval for development of the Rover nuclear rocket. He said, "This gives promise of some day providing a means for even more exciting and ambitious exploration of space, perhaps beyond the moon, perhaps to the very ends of the solar system itself."

THE DIA INVOLVEMENT

The turf wars between intelligence agencies in the early 1960s was intense. Even before the Defense Intelligence Agency was created, the other agencies were highly protective of their sources and information, and reluctant to share power and influence with the other organizations. In 1947, Truman had created Majestic 12 (MJ-12), the Central Intelligence Agency (CIA), and the National Security Council (NSC). Then, in 1952 just before leaving office, Truman formed the National

Security Agency (NSA) as a division of the Department of Defense. By the time President Kennedy took office in January 1961, the various agencies had staked out their own territories and resented intrusion into their affairs. In addition, each branch of the military had its own intelligence capability. The venerable Office of Naval Intelligence (ONI), founded in 1882, was supremely influential and powerful, and probably trumped the youthful CIA in that decade, as did the Federal Bureau of Investigation (FBI).

Further complicating this alphabet stew was the practice of compartmentalization. Secret information was kept contained at the various agencies as well as at different levels within each organization. Consequently, the likelihood that any single high-ranking individual knew what other high-ranking individuals knew was remote. All the intelligence paths met only at the level of MJ-12, and it was this above-top-secret and untouchable committee that pulled all the strings.

In mid-1961, the CIA blamed the failure of the Bay of Pigs operation on Kennedy because he refused to send in air support. U.S. Air Force General Charles Cabell, the deputy director of the CIA at the time, was the most outspoken in placing this blame, although the resentment was also shared by the CIA "foot soldier" participants in the raid. Kennedy,

DIA Headquarters in Washington, D.C.

DIA emblem

on the other hand, blamed the CIA for the botched operation, and then fired Allen Dulles, the longtime director of the CIA, as well as Cabell, and promised to break the CIA up into a thousand pieces. This resulted in unvarnished hatred of Kennedy by the CIA, and was most likely the principal motivating factor in prompting Kennedy to create the Defense Intelligence Agency (DIA) in October 1961. On one level, Kennedy hoped to eliminate military intelligence agency rivalries by combining them. However, in view of his feud with the CIA, it appears likely, in retrospect, that he hoped to have the DIA replace the CIA over time. Ironically, he achieved just the opposite result, as yet another intelligence agency rivalry was spawned. The CIA, it seems, was not so easily deposed. Just as with J. Edgar Hoover, they "knew where all the bodies were buried."

In view of these rivalries, it is safe to conclude that whatever information the DIA obtained from the other intelligence agencies about events prior to its creation was probably given to them only reluctantly. Consequently, whatever intelligence it garnered about alien interactions in the period between 1947 and 1961 was almost certainly incomplete and unreliable, and possibly even included disinformation. DIA operatives had to depend mainly on MJ-12 for that information. So, if MJ-12 decided to withhold what they knew in the interests of compartmentalization, or if they shared any sort of anti-Kennedy bias, the DIA would be left completely uninformed and would be forced to start its operations in 1962 with a clean slate. And that seems to be what happened.

THE KENNEDY DIRECTIVE

Anonymous tells us that President Kennedy issued the directive for the Eben exchange program about six months into the planning for the return visit. That would have been around September 1962. That means that he was probably briefed on the Eben–Los Alamos communications history by MJ-12 around that time. As we noted earlier, the suggestion of an exchange program was first advanced in the fifth message to Serpo sent by Ebel in 1952, at the urging of the alien's military handlers, but presumably ultimately at the prompting of President Eisenhower. So this idea did not originate with Kennedy. In fact, he may have been persuaded to push this agenda by MJ-12. And that explains how the DIA came to be involved in the operation. Certainly, Kennedy would have wanted his new intelligence agency, firmly under the control of himself and Robert McNamara, to take over such a momentous and dangerous operation. It is perfectly understandable that he would have bypassed his old Bay of Pigs adversaries at the CIA to gain this control, even though he had already fired Director Allen Dulles and Deputy Director General Charles Cabell a year earlier. But there may have been a very important, but less obvious reason. There is evidence that Kennedy's distrust of the CIA had deeper roots—that he believed they did not necessarily serve the interests of any administration, but followed more of an independent agenda because they outlasted presidential terms. Although they did answer to congressional oversight, they had such a broad, international overview and so many deep secrets buried in their vaults that they were really powerful enough to do whatever they wanted. Because the DIA was new and was entirely Kennedy's creation, the president could be certain that this amazing story about space travel would not be used in the service of that independent agenda , but would eventually be given to the American people, who had the right to know about it.

But also, from what he knew, Kennedy may have considered the Ebens to be potential enemies at that point, so that makes DIA involvement very logical. After all, the DIA was charged with obtaining enemy intelligence, whether on or off this planet. And it may very well have been their recommendation that convinced Kennedy to agree to the exchange program in order to get as much information as possible about the Ebens' weaponry and motives. If Kennedy did harbor that mistrust, it is further evidence that he was not informed about the Kingman "crash" in 1953, and the subsequent interactions with the Ebens that took place under Eisenhower. That would also explain why none of that information was known by the DIA and released by Anonymous with the Serpo material in 2005. The DIA part of the narrative picks up with the training of the Serpo team in 1963. At that point, the agency was two years old, and had evidently just been brought into the operation by McNamara and Kennedy, so they would have had only those records in their archives.

If this assumption is correct, then Kennedy would have been led to believe that the return visit and the exchange program plan, now under the control of the DIA, constituted our entire agenda with the Ebens, while, in truth, we already had their representatives working with us on the reverse-engineering of their craft at Los Alamos, Area 51, and the Nevada Test Site, under the control of the CIA, the Air Force Office of Special Investigation, the Office of Naval Intelligence, and probably the National Security Agency (NSA). It seems likely that MJ-12 probably bowed to the wishes of the military and the CIA not to allow the DIA to get their hands on that program. They knew that if Kennedy had been given that information, he would almost certainly have complicated matters by trying to involve "his" intelligence organization, and to take over a mature operation, already well advanced over the last nine years. Thus began the era of limited presidential access to top secret affairs. From the point of that first precedent, MJ-12 controlled all interaction with the aliens, and gave the president only the information that, in their judgment, he had a need to know. President Eisenhower was the last president to be fully informed about our dealings with extraterrestrials, and there is reason to believe that even he was not told everything.

So, while the Kingman event was not included in the Serpo story released by Anonymous in 2005, it belongs in this narrative because it fully explains why the Air Force had no reticence about sending twelve Americans to a distant planet for a period of ten years. They were confident of the success of the mission because by the time President Kennedy had given the directive for the exchange program in 1962, the military had already been secretly dealing directly with the Ebens right here on Earth in a reverse-engineering program to duplicate their antigravity discs for nine years. By 1962, the CIA and the NSA had volumes of intelligence about the Ebens. We knew all about their history and their character, and had already established a working diplomatic relationship with their civilization. Furthermore, we were assured of their good intentions because they had given us, as a gift, an antigravity disc to use as a prototype, supported by Eben scientists to help implement our reverse-engineering program. In the absence of those nine years of direct interaction, sending those astronauts on an alien spaceship on a trip of thirty-eight light-years, based solely on some very questionable and shaky communications across the vastness of space, would have been a very risky, one might even say reckless, proposition indeed.

In view of this experience, the CIA and the NSA were perfectly willing to let Kennedy and the DIA take over the exchange program and to incur whatever risks existed. Physically going to their planet was the next logical step in developing a stable exopolitical relationship with the Ebens. They knew that Kennedy would never be able to take the credit for such a grand space triumph and to parade the success of the mission before the public because he would be out of office by the time the astronauts returned, which was expected to be 1975. This, in fact, may explain why the astronauts were sent to Serpo on such an extraordinarily long mission. It appears that this may have been planned so that Kennedy would have been retired from the presidency before they returned, even if he served two terms. They really had him in a straightjacket. Kennedy's silence about the Serpo adventure was critical. If it had been revealed to the public, it might have opened the entire Pandora's box containing all the secrets of our dealings with the Ebens in the reverse-engineering program. They knew how much he wanted a space-based accomplishment, so of course he would want the exchange program. But he had to keep quiet about it while it was in progress because it could have ended in disaster, and that disaster would, of course, have become his legacy as an ex-president if it had been known that it was he who had sent the astronauts to Serpo. It would be said that he had made a monumental error in judgment to have risked the lives of twelve Americans on such a fanciful and hazardous adventure, lacking the intelligence that such a journey required. He would not have had that intelligence because it wasn't shared by the CIA and the NSA. Kennedy knew nothing about what they had learned over the nine-year period since the Kingman event. Yet, because of his passion for his new frontier in space, he was willing to embark on what, to him, was a very high-risk adventure.

So MJ-12 got everything they wanted. They got the go-ahead for the exchange program from a new, inexperienced president who was committed to space travel. And because they did not share their comfort level with the Ebens with him, he believed that it was a very dangerous operation, and that he could not risk the catastrophic blow to his legacy as an ex-president if the twelve astronauts never returned to Earth. So they obtained his silence. They believed that they had manipulated President Kennedy perfectly. However, they left one factor out of the equation. By putting the operation in the hands of the DIA, Kennedy was ensuring that the entire story would eventually be revealed. He made certain that, at its basic level, this organization shared his dedication to government transparency. Kennedy had always been strongly opposed to secret societies and to government secrecy.* He knew that, perhaps not in his lifetime, but in time, this incredible saga of the human journey to another star system would be revealed to the public so that we would begin to understand our place in galactic society. And so, forty-two years after his death, it was.

For more complete information about the DIA, see appendix 8.

---

*In a speech on April 27, 1961, before the American Newspaper Publishers Association, President Kennedy said, "The very word *secrecy* is repugnant in a free and open society, and we are as a people, inherently and historically, opposed to secret societies, to secret oaths, and to secret proceedings. We decided long ago that the dangers of excessive and unwarranted concealment of pertinent facts far outweigh the dangers which are cited to justify it."

PART TWO

PROJECT CRYSTAL KNIGHT

This part of the book draws heavily upon the diary of the Team Commander, who we also know as "102" (see page 97). We are very fortunate to have his personal record of this amazing journey as it was happening. In this diary is recorded far more than simple information and observations. We can also now appreciate all his anxieties, his misgivings, and his impressions. When he says, "I dreamt of Earth" and describes his Colorado home, we experience a wave of sympathy for this man who has volunteered to travel 240 trillion miles from Earth, as we try, with great difficulty, to comprehend what such a journey could be like. We also become spectators to what he is seeing and hearing, and we learn what he thinks about his adventures. We must give many thanks to Anonymous who had access to the actual diary and who painstakingly transcribed it verbatim and sent it in to the Serpo website. In sending in these diary entries, Anonymous prefaced them with a plea to Victor Martinez, the website moderator. He said:

> Attached for your UFO Thread List are just four pages of the Team Commander's diary. The diary contains a large number of pages, all hand written. It took me several days to prepare the attached four pages from the diary. This is the actual, verbatim diary of the Team Commander. It was started the morning of the departure. There were code names for control personnel and three-digit numbers for each Team Member [addressed above]. There are other codes and abbreviations for certain things, which is [*sic*] not explained.
>
> I've typed the EXACT words, phrases, and abbreviations. Nothing has been changed. And likewise, I will ask you to not alter, change, or correct any of the text here as you often do with mine to make it gram-

matically correct. This includes your use of caps to emphasize things I've written; I ask you to not do this with these journal entries, Victor.

In making this plea, Anonymous is communicating to Martinez how important it is to leave the Commander's precise phraseology intact and to reproduce it exactly as he recorded it with incorrect spelling, grammar, and inconsistencies included. He understood that this diary must be given to posterity "warts and all." He knew that someday school children all over the world would read these words in their history books, and they would be right there with the Commander to witness these events as they happened, to see it all through his eyes! So, in the diary extracts in this section, we have tried to maintain this high degree of authenticity and have reproduced the text exactly as it was sent to the website, and we ask the reader's indulgence in reading the diary entries that are laced with errors, some careless, some in the interests of saving time, and some without regard for grammatical accuracy. In some cases, where absolutely necessary, I have added the correct text in brackets. In any case, I am confident that the Commander's intent will easily be understood.

The more we learn about 102, the more admirable he becomes. Having dared to cross the galaxy, he must now shepherd his team through thirteen years of an incredibly challenging existence. They must endure extreme heat, strange food, constant daylight from two suns in the sky, excessive radiation exposure, and little or no recreation. All the while he must carry out his mission to learn all he can about this civilization. There are times when he can easily be mistaken for Captain Kirk from *Star Trek,* such as when he confronts the Ebens for authorizing the cannibalization of his dead teammate for a cloning experiment (see pages 147–51). Interestingly, the team arrived on Serpo within a few months of the date that Captain Kirk first appeared on our television screens in 1966. Apparently, the timing was right for both. Sometimes it seems there is an amazing confluence of reality and fiction.

7

SELECTION AND TRAINING

President Kennedy gave the official directive for the Eben exchange program. The date for the alien landing had been previously set for April 24, 1964, and the landing site was to be at the western border of Holloman Air Force Base, adjacent to the southern entrance to the White Sands Missile Range, in New Mexico. It was originally planned to be only a diplomatic visit during which the aliens would also retrieve the bodies of the nine dead victims of the two New Mexico crashes, as well as the body of Ebe1. But President Kennedy decided to request that the event become an exchange program. This request was communicated to the Eben home planet, and was approved by them around September 1962. As noted earlier, the exchange program idea was originally put forth by Ebe1 in his fifth message in 1952. In a reply directly to him, the Ebens agreed to a return visit, but did not mention an exchange program. They suggested a date ten years in the future. This reply may have suffered from Ebe1's translation and the military handlers at Los Alamos first believed it to be a mistake. But before they could obtain a correction, Ebe1 had died, so that date (1962) remained until the new one in 1964 could be established in 1955 via human to Eben communication. As it turned out, the new date was really twelve years in the future from when Ebe1 had received his reply. Now, in 1962, when communications were much improved, Kennedy realized that the exchange request might now receive an approval, and it did. It was agreed that we would send twelve American astronauts to the alien planet, and they would leave one Eben ambassador here, for a period of ten years. Since the planned date of the landing could not be changed without great difficulty, that meant that the government planners had only about eighteen months in which to select and train our ambassadorial team. It was a very tight schedule for such an unprecedented and complicated program. The selection process alone could easily involve a six-month effort. While President Kennedy had placed the entire program in the hands of the Defense Intelligence Agency, this did not preclude the use of civilian subcontracting agencies. However, it was quickly decided that the Air Force would be the lead agency, taking responsibility for finding twelve volunteers for this historic mission.

The Air Force brought in civilian consultants to help with personnel selection and mission planning. Interestingly, NASA, which became operational on October 1, 1958, had no involvement in the mission. Under the terms of the National Aeronautics and Space Act of July 29, 1958, NASA was to be the official government agency charged with space exploration. And since they had fully absorbed the National Advisory Committee for Aeronautics (NACA), they already had a forty-six-year history of space research that could certainly help to expedite the program planning and preparations. But the NASA charter specified that it remain nonmilitary. So, while the DIA was free to contract with civilian agencies, NASA could not be involved in a DIA program. However, as will be seen below, NASA did participate in training the team.

A GRAND ADVENTURE

The appointed military-civilian selection committee debated for months about the criteria to be applied in choosing the team members. This was precious time lost. It was finally decided that all candidates must be military, but not necessarily from the Air Force. There were to be no civilian members of the team. They had to have chosen the military as a career, and had to have completed at least four years of service. This decision made a lot of sense. The mission required an ultrahigh level of personal and team discipline, since the team members would be facing enormous hardship, and they would have to rely on that discipline to handle the challenges. Furthermore, as previously discussed, Kennedy and the DIA were not absolutely certain that they would not be facing a possible threat from the Ebens, and they wanted people who could defend themselves, and could possibly even find a way to return to Earth by force, if it became necessary. The committee would be looking for individuals with special skills appropriate to their roles in the mission who had additionally been cross-trained in other needed skills. Also, they wanted some evidence that each team member had the ability to step into the shoes of another member, if necessary.

The team members had to be currently unmarried. Presumably, this requirement did not eliminate those previously married, but they could not have any children. Preference would be given to those who were themselves orphans. The goal was to select candidates with as few family ties as possible. Apparently, there was concern that family members might somehow learn of the program and might conceivably make public their worries about its safety. These twelve individuals would be entirely in the hands of another civilization on a planet in a distant star system, and the government had little or no leverage to ensure their safety. Their safety, of course, was highly problematic. This was a very dangerous mission, and the fate of the team was basically unpredictable. If they did all die, we would have no way of knowing whether it was accidental, so we could not retaliate against the Eben in our custody, nor would we want to anger an alien race that had the technology to invade the Earth! This was a delicate balance, but we had enough experience with the Ebens to feel confident that the team had a reasonable chance of survival and returning safely. We basically had learned to trust them, but this was to be the first time any human would be sent out into space beyond Earth's orbit. It was a grand adventure, really more appropriate to the realm of science fiction than to that of reality. But it was worth the risk. If they all or even some of them returned safely to Earth, we would have complete details about a civilization on a distant planet, which would be a window into the universe of incalculable value.

The call for volunteers was placed in military publications. Eventually, after some months, the final twelve—ten men and two women—were chosen. There were eight Air Force selectees, two Army and two Navy. Additionally, four alternates were picked. They would go through the same training as the main team, in case a trainee was eliminated or dropped out. These sixteen were the very best, and most likely to complete the mission successfully.

Relative to selection, the following quoted information came from someone in England who claimed to have been in MI6 (the British intelligence agency) and was involved in the program. It was sent to the website and was not disputed by Anonymous.

> The advertisement which was sent out asked anyone interested in volunteering for a space program [to] apply. It was a semi-classified announcement. The disguise was that the USAF was selecting a special team to travel to the moon and these people must undergo special training and a special selection process. None of the military people trying out for this team knew the real mission. About 500 people applied and that narrowed down to about 160. But there was a problem. Some specialists required on this mission were missing. Besides, the requirement called for each team member to be single, never married, no children and if possible, orphans. The USAF had to go out and recruit two doctors and several other specialists.

SHEEP-DIPPED

The selection committee decided to completely dissolve the identities of all the team members, and to assign them three-digit numbers as their new identities. They wanted to sever all their existing connections to Earth, except those with the mission personnel, and that meant destroying traceable identities. Consequently, they were all "sheep-dipped." According to an article in *Time* magazine on February 3, 2003, titled "The CIA's Secret Army," by Douglas Waller, "If a soldier is assigned highly clandestine work, his records are changed to make it appear as if he resigned from the military or was given civilian status; the process is called sheep dipping, after the practice of bathing sheep before they are sheared." This metaphor is slightly askew. It is really the shearing and not the dipping that applies here. Identities are sheared off like the wool of the sheep. One suggestion was that they all be listed as dead. That was debated and rejected. It was finally decided that they be shown on their official military records as "missing." That seems like a very odd decision, since the "missing" category was really only appropriate in wartime when thousands of military personnel were listed as "missing in action." In 1963, the Vietnam War had not yet begun. It wasn't clear how the military would be able to explain how twelve people could be "missing" in peacetime. How could members of the armed forces suddenly not show up for duty without being labeled as deserters? Stranger still was the decision to either destroy their military records or to place them in a secret storage facility. That meant that no investigator would ever even know that they had been in the military. This means that they could never question the "missing" designation anyway. All other records were to be collected and destroyed or placed in a vault, including Internal Revenue Service (IRS) tax returns, medical records, and any other papers that showed that these twelve individuals had ever existed. That was really an impossible task since there were probably papers such as birth certificates, school and college records, and Social Security cards that may not have been retrievable by the military.

The purpose of all this strict depersonalization is obscure. Very possibly, the committee wanted to prevent the writing of articles and books when the team members returned to Earth and were released to civilian life. The security oath alone should have accomplished that goal. But they weren't taking any chances. By eliminating their credentials, the military made it easy to disavow such writings. They wanted the full report of the space travelers to remain under lock and key for as long as they chose. There could have been a justifiable security reason for this at the time, but now, in retrospect, it seems a great shame that these twelve courageous space pioneers must forever remain in the dark halls of historical anonymity. The final makeup of the team with their new three-digit identities is as follows:

Team Commander	102
Assistant Team Commander	203
Team Pilot #1	225
Team Pilot #2	308
Linguist #1	420
Linguist #2	475
Biologist	518
Scientist #1	633
Scientist #2	661
Doctor #1	700
Doctor #2	754
Security	899

In one of his e-mails, Anonymous answers some questions regarding the Serpo material that were sent in. One question relates to the selection process and offers some additional insight. Following is the question and answer:

> Another question pertained to the team makeup. Why were only 2 females taken? If one considers the monumental problem associated with picking a team of 12 people, where each person must be totally erased from the military system—no family ties, no spouses, and no children—one can see the difficulty that the selection group had. The selection group picked the best team members from a limited pool of military people. The original selection group picked 158 people. The final 12 were selected from that number. If you consider the psychological, medical, and other tests that had to be administered, the final 12 were the best qualified from the original number. Why they chose 2 females was never written. Apparently, those two females were the best qualified in their individual specialty: a doctor and a linguist.

THE FARM

The primary training for the mission took place at Camp Peary, Virginia, on the York River near Williamsburg. This is the not-so-secret main CIA training location popularly known as "The Farm," but officially referred to as the Armed Forces Experimental Training Activity (AFETA). The project was assigned its own training complex within the larger facility, where it could impose its own secrecy and security within the already highly secure CIA system. So it was necessary to go through two levels of screening to gain access to the team training location. Camp Peary was the home training site for the team, but they were also trained at Sheppard Air Force Base in Wichita Falls, Texas, Ellsworth Air Force Base outside of Rapid City, South Dakota, and Dow Air Force Base in Bangor, Maine. They were given high-altitude astronaut training at Tyndall Air Force Base near Panama City, Florida. Anonymous says that they were also sent to unidentified locations in Mexico and Chile for special training.

Camp Peary was a logical choice for training the team, primarily because it was self-contained and it was ultrasecure. Since it was operated by the Defense Department, it was really a military base under the nominal control of the U.S. Navy. It was named after famed naval explorer Rear Admiral Robert E. Peary, who in 1909 was the first man to reach the North Pole. Self-containment was important to the planners of the mission. They didn't want the team to make any connections outside their training facility where they might inadvertently drop a hint or reference to their mission. Camp Peary already had pleasant homes and apartments, recreational facilities, and retail establishments for the CIA trainees that were easily adaptable for the team. Since it was previously a Virginia state forestry and game preserve, the camp residents could even enjoy hunting on the nine-thousand-acre, heavily wooded reservation in whatever spare time they were given. Secrecy and self-containment had always been the major virtues of this facility throughout its history. During World War II, it was the main training camp for the Navy Seabees, and then it was used to house those German prisoners of war, mainly officers, who were supposed to have been killed in action, but had actually been rescued by the Navy. Consequently, the German High Command thought they were dead and couldn't give away any information. The prisoners could live normal lives securely at Camp Peary while they were being interrogated. Eventually, most became naturalized American citizens. The Navy gave the reservation back to Virginia in 1946, but then took it over again in 1951.

The purpose of CIA training at the Farm is to imbue deskbound intelligence operatives with the paramilitary skills they would need in enemy territories. Here, the keenly academic types are "toughened up" and made into pseudo-soldiers. This kind of James Bond commando training was resumed after 9/11 by George W. Bush after a long era of blunders by CIA paramilitary operatives in foreign countries who were in the Special Operations Group (SOG). Waller says in his *Time* article, "Until fairly recently, the CIA, in an effort to clean up a reputation sullied by botched overseas coups and imperial assassination attempts, had shied away from getting its hands dirty. Until about five years ago [1998], it focused instead on gathering intelligence that

Navy Seabees in training at Camp Peary, 1943

could be used by other parts of the government. Before that, traditional CIA officers, often working under cover as U.S. diplomats, got most of their secrets from the embassy cocktail circuit or by bribing foreign officials. Most did not even have weapons training." Now, says Waller, "At Camp Peary, new SOG recruits also hone their paramilitary skills, like sharpshooting with various kinds of weapons, setting up landing zones in remote areas for agency aircraft and attacking enemy sites with a small force." This rebuilding of the Special Operations Group by CIA Director George Tenet required a makeover of the facilities and capabilities of Camp Peary starting in 2001. By the mid-1960s, there had been some improvements as a result of the lessons learned in the Bay of Pigs debacle in 1961. But those facilities were still rather primitive when the Serpo team was in training at the camp between 1963 and 1965.

EXTREME CONFINEMENT

The complete training curriculum, according to Anonymous, is shown in appendix 1. He claims to have received it from an associate named Gene. E-mail moderator Victor Martinez provided Gene's last name—Loscowski.* Anonmyous tells us that Loscowski gave out these details of the training agenda because he wanted to "beef up" the released information with more specificity. Expected to be completed in about six months, it was a very ambitious program. We find out from an anonymous e-mail to the website that it actually lasted about eight months.

The training was intense. Anonymous says, "each team member had to demonstrate their abilities to endure hardship, which included a battery of psychological tests, medical screenings and a PAT (Positive Attitude Test, which is a military test given to pilots and Special Forces personnel) . . . Each team member had to endure extreme psychological and physical training. In one training test, each team member was locked inside a 5' × 7' box buried seven feet underground for five days, with just food and water, no contact with anyone else and in total

*We later learned that his real name is Gene Lakes.

darkness" [see item 14 in appendix 1]. Evidently, candidates with any tendency toward claustrophobia had been weeded out by the selection process. Otherwise, for anyone with such tendencies, this ordeal would definitely have caused the trainee to break down. Apparently, all the team members passed this test. It does seem logical to have imposed a claustrophobic stress test on the team, since they had to be prepared to be cooped up in the alien spacecraft for an extended journey to Zeta Reticuli, and it wasn't known in what sort of living quarters they would be ensconced on Serpo.

TWO LADIES

A section of the Serpo website invited comments from anyone who had direct knowledge about Project Crystal Knight, but preferred to remain anonymous. Many e-mails were received that confirmed all the basic facts about this remarkable journey. Regarding the training at Camp Peary, this letter basically authenticated the story revealed by Anonymous. The writer says:

> I was involved in Project CRYSTAL KNIGHT from about 1960 until 1965. I was assigned as a civilian to this project. I was a CIA employee, with a specialty of survival in a foreign environment. I was a training instructor at the CIA training camp in Virginia. I trained the 12 men—NO women—who went on this mission. They spent about eight months at our training facility. Few knew their exact mission, which was classified "Top Secret/Codeword." I had no other involvement with this mission after 1965. I was very surprised to hear this story come to light now after all these years.

The "NO women" interjection in this e-mail contradicts the claim by Anonymous that two women were part of the original team. There has been other testimony on both sides of this issue, and it remains contentious. It is possible that the twelve trainees referred to here consisted of ten original team members, and two male alternates. The following e-mail helps to clarify the contradiction.

> The Project SERPO information that I just read is NOT totally correct. There were two women in the original 16 selected for training. I helped train the team, including the two women. But after the final selection process—which did not involve any combat training like what was mentioned on the website, www.serpo.org—the two women were dropped from the list. During the training, the team members didn't know their actual assignment. When the final cut was made, the 12 selected were sent to a military prison and then told of the assignment. The 12 were isolated from that point on. The 12 were removed from the payroll of the government and placed in a special file within the Defense Intelligence Agency. The DIA was the controlling agency in Project CRYSTAL KNIGHT, which was the name of the operation.

However, the following e-mail keeps the controversy alive.

> I congratulate and commend you for getting this very important piece of America's hidden history out. I really enjoyed reading this information. I just shared it with many of my old intel friends—and they knew about it! It's truly an AMAZING story, ALL of which is FACT. I guess there still is a controversy on whether two (2) women went or not. At least one (woman) went for sure. I knew six (6) of them while in training as their instructor. Two (2) were NURSES, one was a LINGUIST and I'm not sure about the others with the passage of time.

ESCAPE TO EARTH!

As discussed in previous chapters, we had at least one intact Eben craft in our possession. Anonymous makes mention of the one that crashed in the Plains of San Augustin in western New Mexico. He claims that this craft collided with the Corona disc on July 4, 1947, but limped on westward and finally came down near Datil, but wasn't discovered until 1949. This discovery date was disputed by other Roswell witnesses, but all agree that this disc was virtually undamaged. It was sent to the Wright-Patterson Foreign Technology Division, headed by Dr. Eric Wang, for analysis and reverse engineering. The seriously damaged

Corona craft was also shipped to Wright-Patterson. At that time, Area 51 was not yet in existence. However, after it became operational, both discs ended up there.

We learn from Anonymous that the training plan actually contemplated a possible emergency escape contingency! He says:

> Several selected team members (pilots) were trained on flying an Eben craft, one of which was the one captured near western New Mexico in 1949. The plan called for these selected few to fly the craft back to Earth in case of an emergency. There were four pilots on the team [102, 203, 225, and 308]. These four spent many weeks at the Nevada complex learning to fly the recovered Eben alien craft. It wasn't hard to fly, once one could understand the operation of the controls. I'm sure many of the UFO sightings back in 1964–/'65 around the West could be attributed to these test flights by our team members.

In this reference to the San Augustin craft, Anonymous implies that we had more than one flyable Eben craft in our possession. The other one could only have been the disc that was "delivered" to us at Kingman, Arizona, on May 21, 1953, and was hauled on a tank carrier over the road to the Nevada Test Site (see chapter 5 and plate 6). This is a masked reference to Area 51, which was fully functional by that date, and had now become the major facility for the reverse-engineering and flight testing of all recovered alien craft.*

The belief that our pilots would be able to fly an alien craft forty light-years from Zeta Reticuli back to Earth was a very naive supposition based on lack of knowledge of the scientific principles involved at that time. Apparently, in 1964 nobody in the military-industrial complex had yet conceived the possibility that covering such vast distances could only be achieved by time travel through wormholes. It seems highly unlikely that human pilots would have been able to comprehend that sort of technology in that era. Even *Star Trek,*

---

*See appendix 10 and the testimonies of Robert Lazar, who worked on reverse-engineering alien craft at Area 51 in the 1980s (see www.8newsnow.com/story/3369879/bob-lazar-the-man-behind-area-51), and others.

which debuted two years later in 1966, did not mention time travel, but referred to "warp drives." It wasn't until the advent of *Star Wars* in 1977 that the term *hyperspace,* basically a synonym for the time domain, gained popular currency. Furthermore, the Eben discs in our possession were only small scout craft. They were not meant for long-distance, interstellar journeys. Since we had not yet seen the larger craft when the training took place, it is understandable how this judgment error could have occurred.

TYNDALL TRAINING

Another e-mail testimony confirmed the training at Tyndall Air Force Base in Florida, and the fact that no women were in the final team. Evidently, the Tyndall training took place after the final twelve had been selected.

> My father died in 1995. He was retired from the U.S. Air Force. In 1990, he told me a story about a special mission that he was involved in back in 1965. He told me that this mission was about 12 military astronauts that went to another planet in a spaceship that was found in the New Mexico desert. He said the 12 men were trained at Tyndall Air Force Base, Florida, where he was stationed. He helped train the 12 in space endurance, which he was trained to do. He said the 12 left in 1965 and came back in 1978 and he was there to check them after they returned to this planet, Earth. I didn't know what to think about my dad's story. Back then, I just listened to him and thought maybe he was just making this up. But now I realise [*sic*] he was telling the truth. It is too late for my father to know about this, but I know my father was being truthful to me and that makes me feel good. I look forward to reading more about this incredible story. Since this e-mail describes the last training component for the remaining twelve, I believe that it is safe to conclude that the final team composition consisted entirely of men.

Tyndall AFB was originally an air gunnery training base during World War II, hosting Allied as well as American pilots. One of its famous alumni was Clark Gable, who graduated in 1943. After the war, Tyndall became a general air weapons training facility and officially became an air training command base in 1950. In 1957 it became part of the Air Defense Command (ADC), which is responsible for defending the continental United States and its territories. In 1964, when the Serpo team went there, it had already begun training airmen for high altitude and space operations, and then in 1968, the Air Defense Command officially changed its name to the Aerospace Defense Command.

As mentioned above by one of the trainers, "During the training, the team members didn't know their actual assignment. When the final cut was made, the 12 selected were sent to a military prison and then told of the assignment." One can imagine the reactions of the trainees to this revelation! Certainly, they could not have imagined such an assignment in their wildest dreams, even though some of the training might have suggested some sort of exotic mission. And then,

High-altitude training at Tyndall AFB

Fort Leavenworth, Kansas

after all they had gone through, to be confined at Fort Leavenworth, Kansas, must have added insult to injury. There was, no doubt, an attempt to imbue the team with pride and to stimulate a feeling of the heroism of their mission, and the glory they could anticipate. But they must have nevertheless been shaking in their boots! They knew that the promise of glory was a fiction and all that they could look forward to was tremendous hardship and an early, anonymous death on a distant planet, or a life in long-term isolation to keep them from revealing what they knew. So, it is not surprising that one of the team members "asked to be excused." He changed his mind when he was told that he would have to remain at Fort Leavenworth until the team returned in ten years!

8
THE LANDINGS

In December 1963, Los Alamos received a message from the Eben planet, confirming all the details of the landing. The message specified the time, date, and location that had previously been agreed upon. All the numbers were stated using our location, time, and date protocols. The message told us that two Eben spaceships were already on their way and would arrive on schedule. We learned later that the journey took about ten months, so that means that the Eben spacecraft had been en route from Serpo for about six months of Earth time when the message was received. President Kennedy had been assassinated only a few weeks earlier, and the entire nation was still in mourning at that point. Some of the DIA project coordinators wanted to cancel the exchange program. The fate of the mission was then left to President Lyndon B. Johnson. He was briefed by the mission planners and made the decision to continue with the exchange, although, we are told in a sidenote by Anonymous that the president didn't really believe that it would happen. It is interesting to note here that, apparently, President Kennedy had not informed then–Vice President Johnson about Project Crystal Knight. This is surprising because Johnson had been appointed by Kennedy to be head of the Space Council. Evidently, Kennedy had been told by MJ-12 that the project information could not be shared with Johnson or with the president's Cabinet.

As the landing date approached, the team was ready and idle and probably enjoying some well-deserved rest and recreation, although they remained under surveillance the whole time. Their training had been completed, and they had been given a fifteen-day vacation. In the time just before the April landing date, they were sent back to Fort Leavenworth in Kansas, and were confined in locked cells in the U.S. Disciplinary Barracks and kept under close watch. This reflected the almost fanatical dedication to secrecy by the planning committee. They were simply not taking any chances, no matter how remote, that information about the impending mission might be revealed. One can easily appreciate how depressed the team must have been to have been treated like criminals on the eve of what should have been a grand send-off on a historically momentous and extraordinary journey to the stars! In another time, under a different, less paranoid government, they might have been sent off to the strains of patriotic music, broadcast on international television, and with a cheering crowd in attendance.

THE DIPLOMATIC GREETING

The two alien craft entered our atmosphere on the afternoon of April 24, 1964, right on schedule. These were not scout craft, but were much larger and were considered shuttle craft. The first ship missed the rendezvous point and landed somewhere near Socorro, New Mexico. This was about a hundred miles north of the planned landing site. We sent a message to the craft that it had landed at the wrong place. The second ship picked up the message and made the navigation correction. It landed shortly thereafter at the precise designated location at White Sands, where a greeting party waited. It can be assumed that it was late afternoon at that point, although it could have been nighttime, as was shown in the movie *Close Encounters of the Third Kind*. Since we don't know where Steven Spielberg obtained his information, we don't know how accurate it was. But it was distinctly possible that night had fallen by the time the alien craft had landed and, presumably, the planning committee was prepared for that contingency with appropriate lighting, as in the movie. We are not told what happened to the first craft. More than likely, it also flew to the correct location.

The greeting group consisted of sixteen senior government and military officials. Anonymous does not give the identities of these people,

White Sands Missile Range, New Mexico

but most probably President Johnson was not among them. The twelve team members waited in a bus nearby. Forty-five tons of supplies and equipment stood ready to load onto the alien ship.* A canopy was in place connecting the landing point with the waiting officials. A contingent of Ebens disembarked from the craft and walked under the canopy. Movie cameras and tape recorders were rolling. The Eben official presented us with some technology gifts. Anonymous says, "The Ebens had a crude translator device. It appeared to be some sort of microphone with a readout screen. The senior U.S. official was given one of the devices, and the Eben kept the other one. The officials spoke into the device and the screen showed a printed form of the voice message, both in Eben and in English. It was crude and hard to understand everything that was said." Direct translation was also provided by one of the aliens

*See appendix 2 for a complete list of all supplies and equipment taken by the team. We learn later that the entire cargo was placed on one huge platform, evidently made of wood.

who was designated Ebe2. This was a female who spoke decent English. She later became an invaluable resource on Serpo.

THE YELLOW BOOK

Ebe2 presented us with the Yellow Book. This was a remarkable and generous gift to the people of Earth and clearly demonstrated the aliens' wish to become our galactic friends. About the Yellow Book, Anonymous says:

> It isn't exactly a book. It is a block of material, approximately 2½ inches thick and transparent in nature and appearance. The reader looks at the transparent surface and suddenly words and pictures appear. It is an endless series of historical stories and photographs of our universe, the Eben planet and their former homeworld, and other interesting stories about the universe. It also contains an [a] historical story and various accounts about Earth's history and distant past. . . . I am one of the very few people who has actually SEEN the Yellow Book. . . . As has been commented on by others, it would take a lifetime to read it and another lifetime to understand it.

The Yellow Book also describes the Eben involvement with the evolution of human civilizations on Earth. Evidently, as we learned later, there were some controversial claims made in this regard that caused some recipients of the information to doubt the veracity of the material, which brought up the possibility that these claims were intended to accomplish a hidden diplomatic agenda. On this subject Anonymous says:

> If one reads the Yellow Book and reads between the lines, one would come away with the thought and clear impression that the Ebens had something to do with Jesus Christ or, possibly, Jesus was one of them. Also, if you look at some events that are shown in the Yellow Book (remember, there are no dates shown in the Yellow Book), you can connect some incidents, such as Fatima, etc., with an Eben landing.

We learned later that it was Ebe2 who had translated the book into English.

The Ebens informed us at that time that they had reconsidered the timing of the exchange program, and wished to reschedule it for a later date. They preferred to only retrieve the bodies of their dead compatriots on this trip, and to return to Earth in July 1965 to accomplish the exchange of personnel. This posed an enormous logistical problem for us, since all the equipment and supplies would now have to be warehoused somewhere, and we would have to keep the team motivated and in a highly secure facility for another year. There were also possible political ramifications that could change our willingness to continue with the program if the Johnson administration decided to cancel it. The aliens took the bodies of the nine Ebens who had died in the two Roswell crashes, as well as the body of Ebe1, onboard. We had performed autopsies on some of the bodies.* The remains had been kept at Los Alamos Laboratories in a special state-of-the-art cryogenic facility. The visit lasted about four hours. The film and audio recordings of the entire event have been stored in a vault at Bolling Air Force Base in Washington, D.C.

THE EXCHANGE

The time selected for the return visit was July 16, 1965. It was agreed that this time the landing location would be the northern section of the Nevada Test Site. About this choice, Anonymous says, "Planners did not wish to keep the same location for fear that something might leak." Once again, the extreme security concern becomes evident. The Team members were returned to their lockup at Fort Leavenworth for one month, and then were sent back to Camp Peary to hone their original training and to learn some new skills. This gave them all, but especially the linguists, a chance to improve their ability to understand and speak the Eben language. The linguists were now able to achieve a

*In an interview in 2006 on *Coast to Coast,* Linda Moulton Howe reported on a telephone conversation with a man who had received a firsthand account of the autopsies of two alien bodies in the late 1940s, from a physician who had been in attendance.

passable fluency with the high-pitched singsong speech, but the other team members struggled with the bizarre language.

As before at Camp Peary, the twelve team members remained isolated in their own little community within the larger CIA training facility and did not communicate with anyone other than their trainers. This period coincided with the first year of the Vietnam War in which the CIA Special Operations Group played an important role, so the camp must have been a very busy place while the team was there. In April 1965 they were sent back to "jail" again at Fort Leavenworth to wait out the final three months. By this time, they must have begun to feel like real prisoners, and probably wondered what strange political concerns could have justified such harsh treatment. It is likely that the team morale must have then reached an all-time low, although the mounting excitement and anticipation of the rapidly approaching departure date probably helped to offset their depression.

The two Eben shuttles returned right on schedule on July 16, 1965. This time they landed at the northern section of the Nevada Test Site, as planned. The diplomatic niceties having been attended to in their previous visit, this was strictly a working meeting. The twelve

Three M151 Vietnam-era Jeeps like this one were taken onboard.

The team boards the alien ship
(from Close Encounters of the Third Kind*).*

team members waited in a bus, as before, and the military vans were poised to unload their massive cargo, consisting of 90,500 pounds of supplies, equipment, and vehicles (see appendix 2). I think we can safely assume that there had been extensive communications between Los Alamos and Serpo in the intervening year to refine the arrangements, but Anonymous makes no reference to this. The team boarded the Eben shuttle craft, and the cargo was loaded by military personnel onto one of the craft.

The massive size of this ship can be appreciated when we learn that the entire cargo fit into a single level of the three-level craft! The lone Eben ambassador disembarked from the shuttle and was taken away in a military vehicle. He was then sent to the alien facility at Los Alamos Laboratories.

"SKY KING" TAKES OFF FOR THE STARS

Understandably, the team had no intention of allowing the rigid protocols of the planners to infect their ranks by referring to each other robotically by three-digit numbers. They quickly adopted suitable

Poster for the TV version of Sky King from the 1950s

nicknames for each member, but never used their real names. They did, however, use the "number names" for more formal and written communications. The Team Commander, an Air Force colonel, became "Skipper," the two doctors were "Doc 1" and Doc 2," and the pilots were referred to as "Sky King" and "Flash Gordon." Anonymous doesn't supply the other nicknames. A comment sent to the Serpo website in March 2006 points out an interesting reference that adds authenticity to the entire Serpo story. He reminds us that in the golden days of radio, *Sky King* was a very popular kids' series, along with *The Lone Ranger, The Green Hornet,* and many others. *Sky King* ran on radio from 1946 to 1954, and then a TV version was produced and shown from 1951 to 1959. The TV show reruns were telecast on Saturday afternoons up until 1966. So the show was still on television when the team departed in July 1965. The comment contributor says, "Most people today have either never heard of Sky King or have long forgotten about him. However, it wouldn't be surprising that a young pilot in 1965 would have adopted the nickname of 'Sky King.'" A pilot who was about thirty-five when the mission departed would have been an impressionable teenager in the 1950s, and probably watched the TV show. If he was forty, he probably lay on the living-room carpet like millions of other kids of that era, glued to the radio when Sky King came over the airwaves. Could that young boy ever, in his wildest dreams, have imagined that he would one day be among the first Earthlings to depart the Earth to travel to a distant star system?

The Team Commander kept a diary from the first moment of the mission. Anonymous supplied the following account of that first day from that diary.

Here, for posterity, is the Skipper's exact entry for those first scary moments of that historic mission. Anonymous doesn't explain the acronyms, but we can be quite certain that the M refers to "Mission" in each case. MTC might be "Mission Training Coordinator," and MVC is probably the "Mission Voyage Coordinator," who we find out later does not speak English very well and travels with the team. Consequently, he must be an Eben.

WE GO

Day 1

We are ready. Hard to think we finally made it. Team is motivated and calm. Final briefing by MTC and MTB. Cargo packed in EBE craft. Might have some problems with guns. Will be talking to the MVC. 899 [the security officer] and 203 [the assistant Team Commander] will have overall charge of weapons. No sync system or we don't know about them. Everything moving smoothly. 700 and 754 [the doctors] will give each member final check before boarding. OK, we loaded everything and it fits. But we have to transfer all of it to the bigger ship once we get to rendevous [sic] point. Really excited about this. No reservations by anyone. MTC asked all members to make final decision. The team all said go. We go. Interior of EBE craft is big. There are three levels, this is different than the one we trained on. I think that was a scout craft, this one is a shuttle craft.

*We stored the cargo in lower level. We will sit in the center level and the crew will sit in the upper level. Strange looking walls. They seem to be dimensional. There are three stations, four of us will sit in each station. No seats just benches. We wouldn't fit in those small crew seats. The MVC says we don't need anything special, no O_2 or helmets. Don't know what to do with them. OK, final checks. MTC gave us final words. One pray [*sic*] said. We board the EBE craft. 475 [linguist #2] really nervous. 700 will watch him. The hatch is closed. No windows. We can't see out. Everyone is seated in their respective seats on the bench. No retention harnesses. OK, well, bar across us. The craft is starting engine, or what they call energy thrusters. Seems like we are moving but nothing is happening inside. Still able to write this. Really dizzy now. 102 sitting next to me and he is faint.* Something feels really funny. Have to rewrite this because I can't think straight.*

When the commander says, "We loaded everything and it fits," it may be that the actual loading was done by an Air Force ground crew, since it doesn't seem appropriate that this highly trained team would be expected to perform such arduous labor; that is, loading 90,500 pounds of equipment and supplies, although judging from the way they had been treated thus far, it wouldn't necessarily be out of the question. In fact, that may very well be what happened, since a ground crew would have to have had very high security clearances. The statement "One pray said" was probably meant to be "One prayer said" and would seem to accord with the chapel scene in *Close Encounters* where the clergyman delivers a final prayer to the team, and refers to them as "pilgrims." Since the commander says, "Still able to write this," it appears that, at this stage, his diary was handwritten, although we learn later that the diaries of all the other team members were recorded on cassette tape. Eventually, the commander also reverted to voice recording.

*Obviously, the Team Commander cannot sit next to himself. He means 203. He has gotten the number wrong. Not surprising since he says he can't think straight.

9

THE VOYAGE

In his eleventh and twelfth e-mails, Anonymous sent a verbatim description of the entire trip, written by the Team Commander in his diary. So we have all the details of the team experience from the start of the journey until they landed on Serpo. In this chapter, interspersed with author commentary, we present a firsthand account of this amazing voyage of forty light-years, or 400 trillion kilometers (240 trillion miles), through interstellar space. It lasted only ten months. That means they traveled at about forty times the speed of light! It would be impossible to achieve that velocity with any known means of propulsion, no matter how exotic. The only explanation for this is time travel. As we know from the work of Albert Einstein and Hermann Minkowski, time is the fourth dimension of space, so we must now speak of the space-time continuum. The time dimension is sometimes referred to as the time domain.

The Ebens have obviously developed the technology to travel in the time domain. Apparently, there are portals to this domain at known points in the cosmos. These are now called wormholes (see plate 10). Travel through a traversable wormhole is really travel through time, and is faster than the speed of light. However, it takes time and precise stellar navigation to get to and from the wormholes, and this accounts for the ten months. The Team Commander is evidently speaking of travel through the wormhole when he says in his diary, "We all will feel better once the craft gets out of this time wave, as he calls it [referring to an alien] . . . It was dark but we could make out wavy lines. Maybe some sort of distortion in time. We must be moving faster that light speed, but we can't see anything out the window."

It is very interesting that the commander was able to contemplate the possibility of traveling faster than light speed at that time (1965), since scientifically it was considered impossible based on the Einsteinian determinations. Now, it is not unusual for sci-fi authors and scientists to speak of superluminal speed. For more information about the Eben means of propulsion from a DIA physicist, in reply to two questions sent in to the website, see appendix 6.

A SCREAMING MATCH

The second entry by the Team Commander on Day 1 describes the arrival at the rendezvous craft, which I will now refer to as the "mother ship," and the beginning of the actual star voyage. Mother ship is an appropriate designation, since we learn later that it carried multiple smaller ships.

Day 1—Entry 2

*We made it to the rendevous [*sic*] craft. We don't know where we are, but it seemed like we all fainted or were really confused during this trip. According to my wristwatch, it took about six hours. Or maybe more. We left at 1325 and it is 1939. But not sure of the day. We flew into the big ship. We are standing in a bay or something. There are many EBEs helping us. They seem to understand we are confused. The cargo was offloaded in one big move. The platform containing the cargo was moved without unloading the individual cargo. This ship looks like the inside of a real big building. The ceiling is about one hundred feet high in this area of the ship.*

OK, we are being moved into another part of the ship. OK, we moved to another room or area. What a big ship. I just can't describe how big this is. It took us about fifteen minutes to walk to our area. Seems like it is something special for us. The chairs are bigger. But there are only ten of them. OK, I guess 203 and myself will sit in [a] different location above these seats. We are moved by some sort of elevator, but I can't understand how it worked. Everyone is hungry. We have our backpacks containing some C-Rats and I guess we eat now. But must ask the MVC. I can't find him and we can't communicate with the two EBEs here. They seem to be real nice. 420 [linguist #1] will try to use his language skills. Almost funny. Sounds like a screaming match. We just used sign language indicating we want to eat. One of the EBE brought us a container with something it. Doesn't look good but I think it is their food. Looks like mush or oatmeal. 899 will taste it. Well, 899 said it tastes like paper. Think we will all stick to C-Rats. OK, MVC finally showed up. Told us we will begin trip soon. Two milsl think he means minutes but don't really know. Maybe it wasn't such a good idea to eat before leaving. We don't feel any weightlessness and we don't feel dizzy. But we don't know what to expect from this point on. They are indicating we must sit in the chairs.

We learn from this that the Ebens have obviously developed a type of antigravity technology that allows them to cancel out the weight of massive objects, so that they can be easily pushed into another position. There is no other explanation for how they were able to offload forty-five tons of equipment from the shuttle craft to the mother ship so quickly and effortlessly. The diary tells us that this was done in "one big move" without taking the individual items off the platform, so that means that the platform itself was moved. It's surprising that the Team Commander didn't marvel at that! A platform holding three Jeeps, ten motorcycles, six tractors, eight power generators and much more had to be very large. We learned in the previous chapter in the first entry for Day 1 in the Team Commander's diary that the cargo was loaded onto the Eben shuttle craft by the team itself or by a ground crew. In that entry, the TC said, "OK, we loaded everything and it fits. But we have to transfer all of it to the bigger ship once we get to rendezvous point . . . We stored the cargo in lower level." So it appears that the team loaded all the supplies, equipment, and vehicles directly onto a single, movable platform within the first level of the Eben shuttle craft. Because it was all placed on one platform, the Ebens were able to just move the loaded platform onto the mother ship.

These entries in the Skipper's diary settle once and for all the speculation about how the pyramids of Egypt, Stonehenge, and all the other massive ancient megalithic archaeological sites were built (see plate 11). If the Ebens have this weightless technology, we can reasonably assume that it is commonly used thoughout the galaxy, and that the ancient astronauts all knew about it. The big question now is: Was the secret shared with us? If so, that means that we have probably had this capability since 1965. It absolutely boggles the mind to realize what this could mean to our Earthly industries if this technology was given to them. Certainly we could do away with cranes, and skyscrapers could be built in about half the time. The building materials could be floated up to each floor in minutes. We could conceivably have floating towns and cities. Vehicles would no longer need roadways and could zip along above the streets. If this information lies buried somewhere in Air Force vaults, then we can begin to appreciate the magnitude of the social, technological, and industrial revolutions that will occur once these secrets are revealed.

WE REALLY FELT DIZZY

The team suffered considerable discomfort during the first part of the journey, as described in the Team Commander's Day 2 entry.

Day 2

*I'm not sure just how long we were in the containers. We sat in the chairs and a clear container was placed over us and the chair. We [were] isolated in this bubble or sphere. We could breath [*sic*] OK and could see out, but we really felt dizzy and confused. I think I fell asleep or fainted. I think this is another day, but my watch says one hour since we sat, but I think it is the next day. Our time instruments are located in our backpacks, which are stored in another area of this room. We are still in these spheres but it seems OK. Well, 899 figured out how to get out because he is standing up. He opened my sphere. Not sure if we should be out of* *this thing. 899 said a[n] EBE came in and looked at us and left. Other team members sleeping. 899 and myself walking around this room. I retrieved the time instruments. Seems like we have been traveling about twenty-four hours or so. No windows to see. Originally we were told it would take about 270 of our days. OK, EBE came in and pointed to the chairs, I guess we need to go back and get into them.*

I DREAMT OF EARTH

The following entry in the Team Commander's diary has no day associated with it, and seems to be another description of Day 2, since it gives more details of that first travel day on the mother ship. But, at this point, they are well along in the journey. The Commander is so disoriented that he doesn't realize that he has already written about this first day of travel, and covers much of the same ground. He apparently has no memory of his previous diary entry. In this section, we learn that one man is missing.

I dreamt of earth. I really had some vivid dreams of Colorado, the mountains, the snow and my family. It was if I was really there. I had no worries and never thought of my situation inside the foreign spaceship. Then I awoke. I was confused and disoriented. I was in a bowl, well, it looked like a bowl to me. I don't remember how I got here. My first thought was to my crew. I pushed open the top of this glass bowl and it opened. I heard a hissing sound coming from the seams or seals. I looked around and saw that I was inside a room. Not a room I remember. But all of us were inside these glass bowls. All the other crew members were asleep. I climbed out and realized my legs were really sore. But I climbed out and went to each glass bowl and checked on the crew. I found only eleven of us. Somebody is missing. But who? I am so confused. I'm also very thirsty. I can't find any of those water bottles. We had some but I can't find any. My eyes are really having a problem focusing. But I'm writing this in my

The Team Commander had a vivid dream of his Colorado home.

log, I have to record everything. I found number . . . He is alive. Who is missing[?]

I have to look at each bowl. This room is large. The ceiling looks like a bed mattress. The walls of this room are soft. Not very much in this room, except for the bowls and some tubes running from the bowls to the floor. I see lights flashing on the bottom of each bowl. There are bright lights on in the ceiling. Inside the mattress or something. I can't open these bowls. I've tried everything. I must get some help from the Ebens. I found a door but it won't open.

WE ARE THE SPECIMENS

I can't remember how we opened the other doors. How long have we've been in these bowls? I can't seem to remember much. Maybe the traveling in space causes problems with a person's mind. They told us this during training but we never had anyone travel this far in space before. We are the specimens. Maybe I should go back into the bowl. Maybe I awoke too early. My wristwatch says it is 1800 hours. But what day, what month, what year? How long have I been asleep. The floor seems to be soft with wires running in a criscross pattern. I see some type of television screen in the corner of the room. I think it might monitor the other bowls. I can't read anything on the screen because it is in Eben language. I do make out likes [lines], maybe health monitoring lines on the screen. I hope that means everyone is breathing and is alive. But we are missing one man. Did I forget something? Did someone die? I can't remember. I have some type of rash on my hands. Real burning sensation. Maybe it is radiation burns from something. But where are the radiation monitors we had in our packs? Were [sic] *are our survival packs? I can't find anything. I am returning to the bowl. I am lying down. I will stop making entries in this diary.*

WE HAVE NO IDEA WHAT DAY IT IS

Entry

Since I am not sure of which day it is I won't state a day as to my entry. I'll just say entry. We are all sick. Dizzy, upset stomachs. 700 and 754 gave us medicine to settle our stomachs. But we really feel bad. We seem to be unable to focus our eyes and seem to not know which way to turn up or down and don't know the wasy [way] to sit down. Really bad feeling. Medicine helps a little. We are able to eat a little. 700 and 754 tell us to eat and drink the water we brought along we are doing that and feel a little better can't concentrate on anything so I can't rite anymore right now.

Feel a lot better. EBEs came in and did something to the room. It all seems clearer and we are not so confused and dizzy. We ate again and drink more water. Feeling a lot better. We are out of sphere, but must stay in them at certain times. EBE showed us a series of lights above the entry panel. Green, red, and white lights. If the light is red, we must sit in the sphere. If the light is white, we are OK. EBE never explained the green light. Maybe that isn't good. We have no idea what day it is, only that it is 2319. Our date recorder isn't working very well, according to 633 [scientist #1]. He thinks we have been going for 10 days, but not really sure. We have been confined to this room for the entire time. I think this room was made for us and we are safe in this room. Maybe it wouldn't be wise to leave it. No weightlessness. Don't know how they do it. But we do feel a little lightheaded when we walk. Seems room is pressurized. Ears are popping a lot. If we have to sit in this room for 270 days we will really be bored. We can't really do much, all our equipment is packed away. We have our backpacks but they only contain a few items. We want to clean up, but can't find a bathroom except the containers we are using to relieve ourselves. They are small metal containers that are emptied by the EBE every now and then. EBE brings us food, their food. We tried it and it tastes like paper—really no taste but maybe it is something special for space travel. 700 is eating it. He seems OK but it is upsetting his bowels. Their water is milky-looking but tastes like apples. Strange.

ANTIMATTER PROPULSION

Entry

It has been a long time since I have made an entry. We are guessing we have been on the ship for twenty-five days. But we might be off by about five days. We were locked into our sphere for a long period of time. We had to leave in order to relieve ourselves and finally were able to open the sphere. But when we did, we all got sick, really sick. Dizzy, confused, and some couldn't walk. We had trouble urinating and moving our bowels. 700 and 754, who ate the EBE food, didn't seem to be as sick as us. They treated us with medicine. EBE came in and pointed a bluish light on our heads. We felt better, much better after this. But [he] pointed to the chairs and we figure we must get back into them. We showed him our waste containers and pointed to the chairs in a confused manner. He understood and then left the room. EBE came back with small containers we could place inside the sphere. He also brought in small jugs of the milky liquid and made a motion for us to drink it. So we went back into the sphere and just sat there with the waste containers and the jug of the milky stuff. We drink that and seem to be better, except 518 [the biologist], who seem to be sick. But we were cautioned to stay inside the sphere.

HALFWAY HOME

We learn now that the alien craft exited the time domain at approximately the halfway point. The team members now felt much better and were able to move freely about the ship and to explore and ask questions. Apparently, the team members felt sick and dizzy only when traveling through the time domain. This is important information for NASA and for future manned space probes from Earth. Certainly, our bioscientists will be able to develop some sort of medication to counter this problem. Maybe Mothersill's seasick pills would do the trick. The halfway point in the journey still leaves about twenty light-years, or 200 trillion kilometers, to go. It appears that the Ebens travel through space using very advanced onboard propulsion. We learn here that they may have been using antimatter technology (see plate 12).

Entry

I don't have any idea how long we stayed in that sphere this time. But EBE came in and made a motion for us to come out. We were able to move around without being dizzy or sick. EBE even allowed us to leave the room. We walked along a very narrow hallway for a long period of time, maybe twenty minutes. We then got into some sort of elevator, which moved fast because we could feel the motion. We came out into a very large room that contained many EBENS sitting in seats. Maybe this is the control center. Our escort made a motion to go into the room. We could see control panels which contained many lights. There were four different stations containing six EBENS each. They were in levels. The top level inside this room contained just one seat. One Eben was seated in that chair. We figure he must be the pilot or commander. He seemed busy with an instrument panel. There were many television screens, but they all showed EBEN language and [a] series of lines, both vertical and horizontal. Maybe some sort of graph.

We were able to wander around without any EBEN bothering us. 633 and 661 [scientist #2] were really interested in this. 633 seemed better. There was one window. But we could not see anything. It was dark but we could make out wavy lines. Maybe some sort of distortion in time. We must be moving faster [than] light speed, but we can't see anything out the window.

NEGATIVE MATTER VERSUS POSITIVE MATTER

OK, MVC finally arrived. He explains in broken English, that we are halfway to the home planet. Everything is functioning properly and we all will feel better once the craft gets out of this time wave, as he calls it. MVC says we can walk around any part of the ship, but we must stay together. We must be shown how to operate the movement centers. We are thinking he is referring to the elevators. It seems simple, just placing your hand over one of the operating lights. White and red. White moves it and red stops it. We hear some type of ringing sound, but MVC says it is only space sounds. Whatever he means by that. We were able to walk around the ship, but it is so large it is difficult to understand how such a large ship can move so fast. 633 wants to see the engines. MVC takes four of us to the engine room or whatever they wish to call the room. It contains large, very large metal containers. They are in a circle, with the ends of each pointing into the center. Many pipes or some type of large tubes connects them. In the center of these containers is a coppercolored coil or something looking like a coil. There is a bright light being shined from a point above into the center of the coil. We hear a very dull hum, but no major loud sounds. 661 thinks it is a negative matter versus positive matter system.

We know that 661 was a scientist. His speculation that perhaps the Ebens were using some form of antimatter propulsion at that time was amazing, since that possibility wasn't entertained then, even in classified scientific ranks. This team was well-chosen, and all were apparently highly intelligent and forward thinking. Now we have learned from Robert Lazar, who worked on reverse-engineering alien craft at Area 51 in the 1980s, that the aliens used antimatter reactions to power their craft, utilizing a superheavy substance called Element 115, an element not found on Earth.

Small-scale speculative model of an antimatter reactor on an alien craft. The dome covers the wedge of Element 115. Model designed by Robert Lazar and Ken Wright.

Since Lazar's revelation, scientists at Lawrence Livermore Laboratory, in collaboration with Russian scientists in Dubna, Russia, have been able to create Element 115 by bombarding other more stable elements and have also now created even heavier atoms, Element 116 and Element 118. All these atoms are very unstable and have short half-lives, but are highly radioactive. Element 115 has been assigned the symbol Uup and named Ununpentium. All these elements have now been added to the periodic table.

THE DEATH OF 308

The following entry in the Team Commander's diary was evidently written as they were nearing the end of the journey. Apparently, they all slept through most of the journey. And each time they woke up, they had little or no memory of what had previously transpired. We learn later that 308 (team pilot #2) died of a pulmonary embolism.

> *I am awake again. Ebens are in the room. My bowl is open. Some of my crew are walking around. The Ebens are helping them. I climbed out of my bowl. The English speaking Eben sees me and I ask him if everyone of my crew is alright. He doesn't understand*

"alright." I point to the crew. I say eleven. Where is number twelve? Ebe1 then points to a bowl that is empty and says, earthman is not living. Ok, someone died. But who My crew is walking around in a state of confusion. I can't get anyone's attention. They look like the living dead. What is wrong with them? I asked Ebe1, what is wrong with them. Ebe1 replied, space sick, but will not be sick soon. Ok, that makes sense. I have no idea how long.

We are still flying but don't know how long. Ebe1 brings some fluids and something looking like a biscuit. Fluid tastes like chalk and the biscuit doesn't have any taste. We all eat it and drink the fluid. Almost instantly we feel better. Ok, get organized. Told 203 to round up all crew. Found 308 [team pilot #2] missing. Must be the dead crew member. Ebe1 came back and lead me to 308. He was in bowl, something like a coffin. 700 and 754 will examine 308. Ebe1 cautioned us no[t] to take 308 out. Don't understand the caution. 700 and 754 is [sic] *here. I try to tell Ebe1 that these guys are our doctors and must examine 308. Ebe1 said no, because of infection. I guess 308 must have had some sort of infection and it could be contagious. But is 308 dead? Don't know. We will take Ebe1 advice. 700 and 754 just looked into the bowl and said it looks like 308 is dead. Everyone else looks ok. The fluids and biscuit must have contained some type of energy food. We can focus our eyes and can actually think. No one can remember how we got into this room. All our equipment is here. Everyone is concerned about our status. Ebese are friendly but won't tell us much. 899 is concerned with being locked into a room. 633 and 661 thinks we should be keeping busy. I agree. I order everyone to get their packs and ration belts, inventory everything inside and see if something is missing. That will occupy the team for some time. My wristwatch says it is 0400. But what day? Date? Don't know. Very strange not being able to gauge time. We have no reference inside this room or this spaceship. The year clock that we brought will be unpacked once we can get to the stowed gear. We don't know where that is.*

10

ARRIVAL

The Commander's diary describes the arrival on Serpo.

> *Ebe1 came in. Told us the journey was almost over. Lead us to a hallway. We got into a moving room and moved to another part of the ship. Came out into a large room with many items. I can't identify them, but they look like clothing chests or bedroom chests. We are also lead [*sic*] to a large table with food. Ebe1 tells us to eat. He said good food, eat. We all look at each other. 700 and 754 says lets eat. Ok, we find plates. Seems like ceramic plates real heavy. I chose something that looks like a stew. I then got a biscuit, same as we ate earlier. The drinks were in metal containers. Same fluid we drank earlier. We all ate. Very little taste in the stew. Something like potatoes, maybe cucumbers, some type of stems. No[t] really bad. The biscuits taste the same. Everyone sat and ate. We found something like apples but didn't taste like apples. Sweet and soft, I ate one. Leaves an aftertaste in my mouth. Team looks happy. Some are joking about not having any ice cream. Ok, the MVC is in room. First time we saw him. He speaks through Ebe1. The language really bothers my ears. The high pitch sounds and then the vocal tones sound very strange. Ebe1 tells us MVC wants us to prepare for landing [see plate 13]. Ok, how do we do that. We must go into bowl room and get in bowl. No one wants to do that but if we have to we will do that. We are lead back to moving room and travel back to bowl room. We climb into the bowl. Some*

use pot for relief of our bowels and bladders. Then climb into bowl. The lids close but we are awake. Just lie there. I fall asleep.

WE SEE TWO SUNS

*The bowl lids have opened. My wristwatch says 1100. I guess it is still day 1. We climb out. Ebe1 is at our side. He tells us, Landing home. Ok, I guess we are there. We gather our gear. 700 reminds us to wear out [*sic*] sunglasses once we exit. We pack up and walk down long hallway. Then into another moving room. We travel for one minute. Then door open. We are in a large room. We see our stowed gear. Many smaller spaceships are stored here. A large door opens. Bright light. We see the planet for the first time. We walk down the ramp. Large number of ebens waiting for us. We see a large eben, largest one we have seen yet. He comes forward and starts speaking to us. Ebe1 translate[s] a welcome message from the leader. I guess this guy is the leader. About one foot taller than the others. The leader tells us we are welcome to planet, he called it something we do not understand. Ebe1 isn't doing a good job translating. But we are lead to an open arena. Looks like a parade field. The ground is dirt. Looking up, I see blue skies. The sky is very clear. We see two suns [see plate 14]. One brighter than the other. The landscape looks like a desert, Arizona or new Mexico. No vegetation that we can see. There are rolling hills but nothing but dirt. This must be the central village or town. We landed in an open area with large structures like electrical towers. Something is sitting on top of these towers.*

THE MIRRORED TOWER

On the center of the village is a large tower. Looks like a concrete structure. Very large, maybe three hundred feet. Looks like a mirror is placed on top of this tower. All the buildings looks like adobe or mud huts. Some are larger than others. Looking in one direction, can't tell any actual compass readings, but there is a very large structure. All the ebens are dressed in the same clothing, except for some of the ebens who were on the spaceship. Now I see others dressed in a dark blue outfit, different from the others. Each eben is wearing some type of box on their belts. All have belts. Can't see any children, but maybe they are the same size too. Our boots leave an imprint on the soil. The brightness is almost too much for our eyes without sunglasses. Looking around 360 degrees, I see buildings and barren land. Can't see any vegetation. I wonder where they grow their food. What a planet. Hard to believe we will have to live 10 years here. But a journey of 1000 miles begins with one foot step. Can't remember who said that but that just came to mind.

Interestingly, the Ebens offloaded the supplies and equipment onto sixteen separate pallets. The commander was not able to view the offloading procedure, but apparently it was accomplished very quickly so it seems likely that the Ebens manually transferred everything to sixteen pallets and then floated the pallets into the underground storage area. Perhaps they were able to make the heavy items weightless when they moved them onto the pallets.

Serpo is a rather desolate-looking planet, and the commander is clearly not very happy about what he sees. In his words, "What a planet. Hard to believe we will have to live ten years here." We later learn that this is not the Ebens' native planet, but was chosen as a refuge after the volcanic devastation of their home planet.

WELCOME TO THE PLANET SERPO

But large numbers of ebens are welcoming us. They seem friendly. Then, almost shocking us, someone speaks English. We all look around and see an eben. This eben speaks very good English. This eben, we call ebe2, speaks almost fluent English, with exception of not really pronouncing the letter w. But ebe2 does a good job speaking English. Ebe2 says we are welcome to the Planet Serpo. Ok, that is the name of their planet. Ebe2 shows us a device and tells us every one of us must wear it. It looks like a small transistor radio. We put it on our belts. The weather is extremely hot. Asked 633 to take temperature. 633 says it is 107 degrees. Very warm. We take off top jacket and [that] leave[s] us with one piece flight suit.

THEY ALL LOOK ALIKE

The Ebens looks at us but seem very friendly. Some are wearing some type of shawls. I asked Ebe2 about that, Ebe2 says they are females. Ok, I see. They all look alike. Really hard to tell one apart except for uniform. Some have different colored uniforms. I asked Ebe2 about that, Ebe2 says military uniform. Ok, that makes sense. Ebe2 lead[s] us to a series of huts, looking like adobe style houses. There are four. Behind them is an underground room, or storage area. It is built into the ground, underground. We have to walk down a ramp. The doors look like a military igloos that store our atomic bombs on Earth. All our gear taken off the spaceship is stored there. We walked down into this area. Very large room. Very cool, a lot cooler. We might have to sleep here. All our gear is there. Sixteen palates [pallets] of gear. This igloo is made up of something like concrete but not the same texture. Feels like soft rubber but still hard. The floor is made of the same stuff. There are lights in the ceiling. Looks like spotlights. They have electricity. We must inventory all our gear sometime. We go back to huts. The huts are cooler than outside. But still very warm. We must get organized. I tell Ebe2, we will need to be alone and get organized. Ebe2, then I realize Ebe2 is female, says ok. We will be left alone. I asked for the body of 308. Ebe2 seems to be confused and doesn't know about any body. I explained to Ebe2 held her hands across her body and bowed her head. It was really an emotional site because she was almost crying. Ebe2 told us that the body would be brought to us but she must check with her trainer. The word trainer kind of shocked me. Is Ebe2 in training and someone is teaching her? Or is the word trainer in Eben something different in English Maybe that means leader or commander. Don't know.

But Ebe2 left. I told 203 to gather everyone in the lower storage area. We will have a team meeting. 633 suggested we start the calendar as of today. It is 1300 on our Day 1 on planet serpo.

The team was able to use all their electrical appliances. They plugged them all into a sealed black box, and they all worked. It is believed that the team was able to use the Eben device that they had developed to communicate with Earth. There didn't seem to be anything to prevent this, except for the slim possibility that it only worked with the Eben language. But even in that case, the Ebens would have translated messages into English. For some reason, Anonymous never discusses the utilization of this communications device by the team.

It is interesting to note that the Ebens have electricity, but no air conditioning! This is one of the many paradoxes of the Eben technology and civilization. They are advanced and civilized, and yet they are primitive and simple at the same time.

The female alien that the Team Commander designated as Ebe2, who speaks excellent English, is clearly the same Ebe2 who had translated the Yellow Book, and who presented it to our greeting party at the first landing in April 1964, and who also translated the conversation at that time. It is coincidental that she was given the same name both times because none of the team members had been present during the greeting at the April landing. They were all waiting in a bus. Her emotional reaction in this bit of drama regarding the body of 308 reveals a sympathetic and compassionate nature, which the team learned was generally characteristic of this race. As we will see, this first incident about the recovery of 308's body was the opening round of what turned out to almost become a major diplomatic problem very early in the game. In this next diary entry, we see that Ebe2 becomes a very important link between the team and the other Ebens.

MAYBE THEY HAVE BOOKS ABOUT EARTH

We had a serious problem. How do we explain our science to an alien entity, who doesn't know Einstein, Kepler, or any of the other scientists of our time. Simple mathematics seems so foreign to them. Ebe2 is the smart one. She seems to understand our language more than 1 and 2. She even seems to understand our basic math. We started with the basic math. 2 plus 2. Then progressed on. She understood and even caught on so quickly that she continued on without our help. We realize she has a great IQ when she repeated 1000 times 1000 and came up with an answer. We showed her our slide rule. It took her a couple of minutes to figure it out, although I don't think she fully understood all the symbols on the slide rule. She is really something. We found a personality in her. Maybe because we have more contact with her than the others. She is very warm hearted, one can just sense that in her. She really cares about us and she even worries about us. During our first night, she seemed to make sure everything was just right for us. She warned us about of the heat and the light. She mentioned that SERPO does not get dark, like earth. I wonder how she knew that? Has she visited Earth? Maybe she is just educated on earth's properties. Maybe they have books about Earth. Anyway, on the first night, she told us about the winds, what winds. Gusty winds start just when the one sun sets. The other sun does not but stays in the lower horizon. The winds blow dust, free dust into our huts. We had a very difficult first night. We call it night but it seems that the Ebens just call it a period of their day. Ebe3 [Ebe2] knew the word day but didn't compare it to earth's day. Maybe she hasn't been to Earth. We didn't sleep well during our night. The Eben's don't sleep, like we do. They seem to rest for periods of time and then wake and go about their business, whatever that might be. When we awoke, Ebe2 was there, outside our hut. I opened the door and she was waiting. Why? How did she know we were awake? Maybe the huts are being monitored by some sensor. Ebe2 told us to follow her to the eating place. She didn't use the word dining hall or chow hall or facility. She used the word place.

The fact that the team used a slide rule for computations really dates this entry, and authenticates the time frame. The introduction of

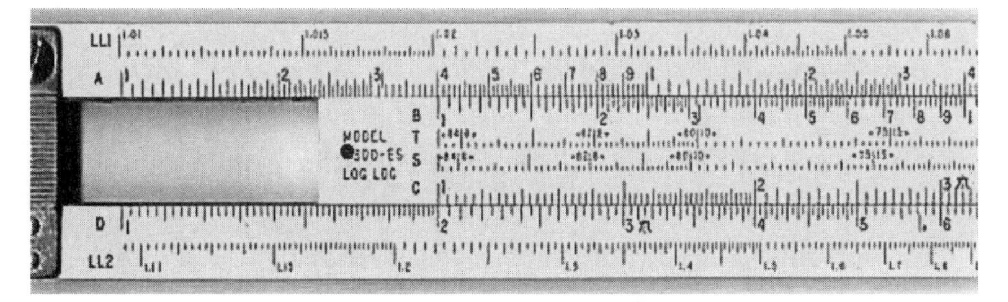

A slide rule still in use in the 1960s

inexpensive scientific calculators in 1974 made the slide rule virtually obsolete. But in the 1960s, all engineers routinely carried slide rules on their belts. The fact that Ebe2 was able to quickly understand the use of the slide rule with its cryptic symbols, tells us a lot about the high Eben intelligence.

The Commander speculates that perhaps Ebe2 had visited Earth. At this point, he did not know that Ebe2 had indeed visited Earth in April 1964. And evidently, he knew nothing about the Yellow Book, or he would have been informed that she had been the translator. The fact that MJ-12 kept this very important piece of knowledge from the team is astonishing. In this case, the paranoia and secrecy went too far. This team had given up a great deal to make this journey and they deserved to know everything possible about the planet that would be their home for the next ten years. Someone should have briefed them about the Yellow Book.

WE ARE THE ALIENS

*Once I gathered the team, we walked across the village, I'll call it a village for the sake of wording. We entered a large building. It seemed to be large, based on the small stature of the Ebens. There were [*sic*] food on tables. I guess we'd call this place a chow hall. Ebens looked at us but carried on eating. Do they not cook inside their huts? Maybe everyone eats here. We walked to the food tables. Same food we saw and ate on the spaceship except some items were different. They bad [had] large bowls of something like* *fruits. Strange looking things. They also had something like cottage cheese, it tasted like sour milk but after the initial taste, alright. I encouraged each team member to eat and drink. We might as well get use to the food. But 700 tells us to eat just one of their meals a day and stick to our C-Rations for the other meals. That way, our system will adjust to the Eben food. We sat at a table, small compared to our standards and ate. The Ebens, numbering probably about 100, just ate and didn't really bother us. Every now and then, we would catch Ebens staring at us. But we were the freaks, not them. We are the visitors. We are the aliens. We must really look strange to them. We all look different, they all look the same. How can they compare us to them? They can't. We stare at them, they stare at us. We then see a different looking Eben. Very strange looking creature, large, long arms, almost floats along with long legs. Can't be an Eben. We all stare. This creature just floats by us and didn't even look at us. I find Ebe2. She is eating with three others. Once I approach her, she stands and almost bows her head towards me. Maybe just a greeting, have to remember that. I ask her abet [about] the creature we saw, I ask if that was some other type of Eben. Ebe2 seems confused. She asked me what creature. I used the word creature. Maybe that was an insult or maybe she didn't know the word. I pointed to the thing at the other end of the building. She then saw what I meant. Ebe2 said, no Eben, just visitor. Like you, pointing to me. Ok, I see, they have other alien visitors here. I guess we are not the only ones. I then asked Ebe2 what planet did that visitor come from? Ebe2 said something like CORTA, not sure of the exact work [word] although I asked her to repeat it twice. Ok, where is CORTA? She walks me to a television, at least it looks like one. It is positioned in the corner of the building. It is set up like some type of command station. She places her finger on the glass and something appears.* The universe? Star*

*This technology is now commonly used by television anchors, commentators, and meteorologists. They routinely manipulate images on the television monitors using their fingers, and bring up statistical underpinnings. And it is used in the iPad and all the new tablets. Quite possibly, the Ebens helped us to develop this capability.

systems I don't recognize any of them. She points to a spot and says CORTA. Ok, where is Earth? She points to another spot and says earth. Based on this television glass space map, CORTA and Earth are very close. But I don't know the scale of this map. Maybe they are a trillion miles apart or maybe 10 light-years (see plate 15). But they seem close. I'll have to have one of the scientists look at this and see. Ok, I thank Ebe2. She seems pleased. She almost looks like a[n] angel. She seems so very nice. She touched my hand and pointed to my table and said, eat. Good eat? I laughed and said, Yes, good chow hall food. She looked puzzled. I guess she doesn't know what a chow hall is. I pointed to the building and said, chow hall, earth eating place. She repeated what I said, chow hall, earth eating place. I laughed and walked away. Now she will think all earth restaurants are called chow halls.

The description of the creature from CORTA sounds very much like the alien we now call the Mantis because it resembles a praying mantis. Many contactees have reported seeing these aliens on spaceships when they were abducted. They are invariably described as kind and compassionate beings. The fact that aliens from other star systems are permitted to freely mingle with the populace testifies to the galactic cosmopolitan nature of Serpo. In the chow hall incident, we see that a very affectionate, trusting relationship has sprung up between the Commander and Ebe2, who has evidently been designated as their combination translator, guide, and hostess.

11

ADJUSTMENT

The Team Commander's record of the first day on Serpo continues:

We retuned to our huts. We must get better organized. We have a meeting. Everyone seems ok. We wonder about latrines. Where do we relieve ourselves Ebe1 comes by, almost as if he was reading our minds, maybe they can even do that. He tells shows [sic] *us pot in hut. We all wondered what that was. Ok, that is our latrine. Realized that won't work very well but we'll do what we can. Then we realize the pot has some sort of chemical inside. Our wastes are dissolved or something like that. Can't really tell. Each of the four huts has one. That will work for the time being.*

MUST BE OVER 140 DEGREES

Ebe2 [Ebe1] tells us to walk on ground. Not sure what he means by that but 420 says that might we [be] to just walk around. OK, that we will do. I organize the teams 102 will remain with 225. I want 633 and 661 to look at the television glass map and see if they can tell which star system is CORTA. I ask 518 [the biologist] to take temperature readings and general weather observations. I know it is hot. Very hot. Most [Must] be over 140 degrees. 754 warns us to keep covered from sun radiation. 754 says radiation levels are high. Doesn't sound very good to me. This reminds me of Nevada. 1956, during one of the atomic bomb testing. We had

> *hot weather and we had to worry about the radiation from the atomic blasts. Now we are on a strange planet 40 light years from Earth and we have radiation and heat. But we have to explore—that is why we were sent. We starting walking around 475 will take photographs with our military cameras. I hope the film is [not] affected by the radiation. How do we develop it? Maybe we didn't think of everything. I team up with 225. We walk to a large open door building. We walk in and it looks like a classroom but no Ebens are here. There is a large television tube. It takes up the entire wall. There are some lights flashing on this television tube. We examine the tube. It is very thin. I wonder how it works. Where are the tubes or the electronics. But maybe they are more advanced in this area than us. They must be. We don't find [or] touch [anything] else in this building. We move on. wow, it is so very hot, I hope I get use[d] to it.*

The commander's comparison of conditions on Serpo with the 1956 atomic bomb testing conditions in Nevada, especially with regard to radiation exposure, tells us that he must have been involved in those tests. This is the second piece of information revealed about the Commander's past. We previously learned that he was from Colorado. The huge, thin television "tube" without electronics showing sounds suspiciously like our modern flat-screen televisions. Once again, was this technology given to us by the Ebens? It's interesting that the Commander is surprised that "They are more advanced in this area than us." Why would he be surprised that they are ahead of us in television technology when he knows that they have mastered time travel?

The Commander continues:

> *We find a large tower. Looks like a antenna tower but it has a large mirror. We saw this when we first arrived yesterday. We find an Eben standing near the door but he moves to one side. We ask him if he understands English. He just stares at us, but seems friendly. I guess he doesn't speak English. We move on into the building. Can't find any stairs. But we see some type of round*

glass room. Maybe it is an elevator. Then we hear English. We turn around and find Ebe2 standing there. Where did she come from? I [ask] her if we may explore this building. She said yes, of course, She points to the glass room and says go up. Ok, we enter the glass room. The glass door closes and go[es] up real fast. I[n] no time we are at the top.

THE SUNDIAL TOWER

But what is this. We ask Ebe2, what is this. She points to the sun and then points to the top of the room, where the mirror is located. She then points to the ground. Ok, we see it. The tower is in the middle of a circle. The circle is situated on the ground. In each quadrant of the circle is a symbol. I see that the sun is directed through the mirror maybe this isn't a mirror as we know it since the sunlight travels through it, but once the sunlight travels through it, the light is directed on a symbol within the circle. Ebe2 says that when light contacts symbol, then Ebens make change. Not sure what that means. Maybe she means it tells the Ebens what to do. 225 seems to think it is a sundial. When the sun strikes a symbol, the Ebens change what they are doing and do something else. Maybe the Eben day is structured. Or maybe this is their clock. Strange. But we are on an alien planet. I'm glad I still have my sense of humor.

WE HAVE TO START USING EBEN TIME

This is only our first day, first day of school. We have a lot to learn. We have to keep an open mind. We can't keep comparing things to Earth. We have to open our minds to new ideas and new science. All these things are foreign to us but we must learn. I pointed to my wristwatch and then pointed to the ground and in a gesture to show Ebe2 that the two items are timepieces. Not sure if she understood. But I said time and she then understood. Yes, she said, Eben time, pointing to the ground. I pointed again to the wristwatch and said, earth time. Ebe2 then almost smiled and said, no, no earth time on SERPO. Ok, that makes sense. 225 said, she just told us that earth time doesn't work on SERPO. Yes, I guess she did. What good is our watches or time device[s] here? They don't work. We have to start using Eben time. But we must also maintain our time because we must know when to leave. Ten years seem like a million. Maybe based on Eben time that it might just be a million years. Lets hope not. No time to think about home. We have a mission and duties to perform. We are a military team and we must maintain that idea. 225 and myself go back into the glass bowl and go down to the ground.

The sketch shown here was made by the Team Commander to attempt to capture his view of the base of the mirrored tower. He considered this important enough to draw because the sun's focus on these objects dictated the day's activities for everyone in the community, and each community had such a tower. See page 142 for what Anonymous says about the Commander's sketch. It should be noted that Anonymous claims the sketch was made in 1967. Since the team arrived on Serpo

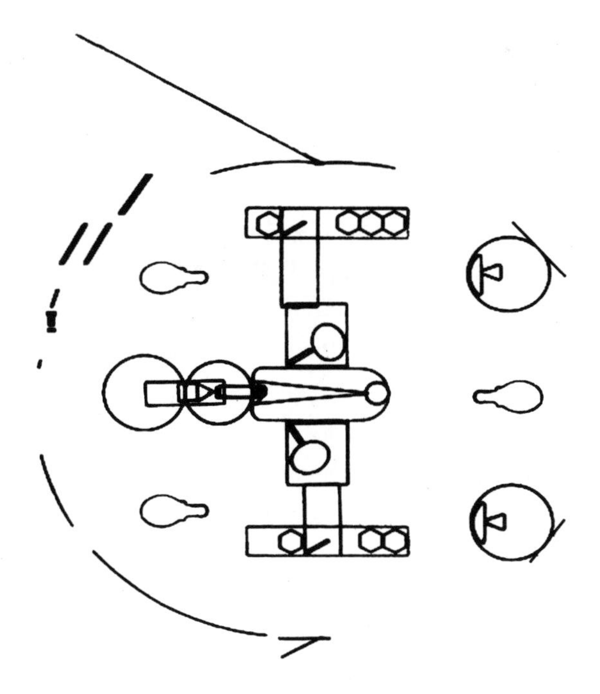

in mid-1966, obviously this sketch was not made on the first day on the planet, but a few months later. The words in brackets were added by e-mail recipient and editor Victor Martinez. Anonymous mentions drafting templates to explain that these templates were used to make the sketch. It is not a freehand drawing. Website moderator Bill Ryan adds the following information about how it was drawn: "It has since been shown conclusively that this drawing was made with a Berol RapiDesign R-22 architect's drawing stencil. However, it is also known that this very common template was on the market a number of years before the Serpo mission departed in 1965, and may well have been included among the equipment taken by the Serpo team."

COMMENTS BY ANONYMOUS ABOUT THE SKETCH

A great deal of planning went into the 10-year mission to Serpo. Several teams of officials planned what equipment to take. These officials tried to imagine every conceivable situation where certain types of tools and other equipment might be needed. One area was capturing the view and makeup of objects, artifacts and landscapes on Planet SERPO. The mission team took cameras—sixteen different types of cameras and drawing/drafting equipment. Although no Team Member was trained in drafting, three Team Members had drawing experience from their college days. Several different types of drafting templates were taken on the trip. The planning officials tried to envision every type of situation where a template might be needed.

The drawing, made by the Team Commander, depicts the base of the sundial. Each object meant a certain time of day to the Ebens. When the sun was directed—through the sundial—onto the object at the base of the sundial, that meant a specific task [or change of] to the Ebens. For example, it might signal a change in work schedules, a time to rest, a time to eat, a time to celebrate, etc. After a few years, the team learned each symbol and learned the meaning of each symbol.

This drawing is a copy of the actual drawing made by the Team Commander in 1967. The Team also took many photographs of the sundial symbols . . .

The regimentation of Eben society is surprising. Our team was able to determine that the average IQ of the Ebens was 165 on our scale. It wouldn't be expected that aliens with such high intelligence would subject themselves to such rigid control over their daily activity. On Earth, we associate creativity with complete freedom of expression, and we understand that creative people must be permitted to live perhaps eccentric lives. We actually expect writers and artists to not conform to conventional patterns of life. Perhaps this is the key. The Ebens did not display the sort of individuality that we have, and consequently there was little or no creativity in the Eben race. They are not very innovative. They simply accept the precedents and follow the rules handed down to them from earlier generations. Intelligence alone does not guarantee creativity. From this, we can begin to appreciate the unique qualities of the human soul and the importance of diversity.

EBEN FOOD

*We walk into another building, large again. Inside we find rolls and rolls [*sic*] of plants. This must be some sort of greenhouse. They are growing food. Many Ebens are inside. They look at us with some glares. But we just walk in and around. One Eben comes up to us and speaks but in Eben. He seems to be telling us something. He points to the ceiling and then to our heads. Maybe he is telling us to cover our heads. We must find Ebe2. We go back outside and then finds [*sic*] Ebe2. She always seems to be near. Now we understand why: she is monitoring us by the devices we are wearing on our belts. I ask Ebe2 what is that building. She says foodmaking place. Ok, maybe we contaminated the place. We told her that another Eben spoke to us and then pointed to our heads. Ebe2 seemed confused and walked back into the building with us. The same Eben exchanged words with Ebe2. Ebe2 then told us that we must wear a cover on our heads in order to enter. Why? But we don't argue. This other Eben produces some sort of cloth had [hat] and we wear it. We walk about, the Eben seemed happy. We look at their plants. They are growing food in soil. They* *have a watering system. They also have some sort of transparent cloth over each plant. I point to the watering system and ask Ebe2 if that is water, drinking water. Ebe2 said yes. Then she senses we are thirsty. Ebe2 leads us to a place near another entrance and offers us water, at least we think it is water. Tastes like chemicals but it is water. Actually tastes good.*

This is another example of Eben technology that seemed novel to our people at that time in the mid-1960s, but hydroponics is now commonly used here to grow vegetables. It permits complete inexpensive control over plant growth, and requires no soil. It is beginning to appear that we learned a great deal from our Serpo experience. In his sixth information release, Anonymous gives the following details about the Eben food, taken from the three-thousand-page debriefing document that he had in his possession.

Food . . . was a problem for our team members. Our team consumed mostly military-style C-rations, but eventually had to switch to Eben food. The Ebens had a variety of food items. They grew vegatables [*sic*]. Our Team found items similar to potatoes, but they tasted different. They had some type of lettuce, turnips, and tomatoes. They were the only items similar to ours. The Ebens had other vegatables [*sic*] grown. These were strange looking round items with long vines. The Ebens cooked the vines and ate the large portion of the plant raw. The Ebens had some type of white liquid, which we first thought was a form of milk. But after tasting it, our Team realized it was different, both in taste and content. The liquid came from a small tree located in the northern portion of the planet. The Ebens literally milked the tree for the liquid. It appeared to be some sort of pleasure to drink the stuff. Our team members never got a real "taste" for the liquid.

The Ebens cooked food. They make pots of stew, which was extremely tasteless to our team. We used a lot of salt and pepper. They also baked a form of bread. It was nonyeast bread and tasted fairly good, but caused extreme constipation to our Team Members. We had to drink large quantities of water in order to digest the bread. The one

U.S. hydroponics farm

common food that Ebens and our team members liked was the fruit. The Ebens ate a great quantity of fruit. The fruit, although different from anything we saw, was sweet. Some of the fruit tasted something like mellons [melons], while others tasted like apples. Another problem was water. The water on Serpo contained a number of unknown chemicals found by our team. Our Team eventually had to boil the water before drinking it. Seeing this, the Ebens built a large plant that processed water for our Team.

EBEN ELIMINATION

Anonymous continues:

The Ebens did not have a physiological need to release body wastes as we did. The Ebens had small collection locations in the residence for their body wastes. But the Eben's body was extremely efficient in processing all food taken in. Their body wastes consisted of a small amount of fecal matter, similar to a small cat dropping. Our team members never saw any urine excretion from an Eben. On the other hand, our team members' wastes consisted of bulk quantity of both fecal matter and urine. The Ebens had to dig large waste reception sites for our 12 team member's waste. The Ebens accommodated our team.

12
CONFRONTATION

And fourth (but this would be a long term project, which it would take generations of totalitarian control to bring to a successful conclusion) a foolproof system of eugenics, designed to standardize the human product and so to facilitate the task of managers.

FOREWORD TO ALDOUS HUXLEY'S
BRAVE NEW WORLD (1946)

Following the death of 308 on the trip to Serpo, the Ebens took control of the body without explanation. Upon arrival, the Team Commander requested the body of 308, and was told that was not possible. In the following diary entry, the Commander relates how his attempt to retrieve the body escalated into a tense confrontation and how Ebe2 tried desperately to defuse the situation.

THEY TOOK ALL 308'S BLOOD

The leader of the Ebens is a larger creature than the others. He seems to be more aggressive than the other Ebens. When I write aggressive, I don't me[an] in a hostile way. He seems to be the boss, similar to me, as the Team Commander. His voice, although after all this time I still can't understand any words, is harsh and with a tone that is different from the rest. 203 claims the leader *has an attitude. I agree. He is extremely friendly to us and has accommodated all our requests. The leader has requested many things from us. Most of which we have provided. One strange thing is our blood. He wanted all of us to supply a blood samples [*sic*]. Ebe2 explained that the blood or health fluid, as Ebe2 explained it, was necessary for them to supply any medicine to us if we should ever need it. 700 and 754 feels [*sic*] that a blood sample might just be used for other purposes. We have allowed the Ebens to utilize 308's body for experiments. They took all 308's blood—that was without my approval. I wrote about that in log 3888. We had a very tense situation with the Ebens about that. When we traveled to the building housing 308's body, we were confronted by several Ebens. Ebe1 [This Ebe1 is the Eben leader referred to above. He is not the same as the Ebe1 who crashed at Roswell] showed up and I explained that we wanted the body of 308. Ebe1 told us the body was in storage and we could not take it. We told Ebe1 that we will take it. The eleven of us walked by the Ebens, numbering six and walked into the building. They did not attempt to stop us. While inside, we could not open any of the containers. There was some type of system, maybe crypto style, that was being used to lock the containers. We did find the container housing 308's body. We decided to send 899 to our storage area and get some explosives to open the container. Ebe2 showed up with the leader. Ebe2 was extremely polite and asked us to wait. She used the word please several times. In fact, she actually used the English word beg.*

We backed off and I told Ebe2 that we wanted our friends body and we wanted to examine it. Ebe2 translated that to the leader. There was a long exchange of words between the two. Finally, Ebe2, who seemed very frustrated told us that the leader would like us to go to another location and speak with another Eben, a doctor, about the body of 308. Ebe2 explained that everything [we] wanted to know about 308's body would be explained to us by the Eben doctor, who Ebe2 said spoke English. I told Ebe2 that I would leave 899 and 754 here to guard the body and I would take the others to the location containing the doctor. Ebe2 translated that to the leader. Again, there was a long, drawnout exchange of words between the two. It lasted for several minutes. Finally, Ebe2 stated that the leader would like all of us to leave this building and go to visit with the doctor. I told Ebe2, no, I would not leave the body of 308 alone. I felt that this was going to be a confrontation. I told 518 and 420 to go back, obtain our handguns and come back ASAP. I was not going to allow the Ebens to counterman[d] my decisions. When Ebe2 heard this, she told me to wait and placed her hand against my chest. I told her to translate that to the leader. Again, there were several minutes of word exchanges between the two. Ebe2 then stated that the leader would bring the doctor here to discuss the situation with us. Ebe2 asked me to please don't send your men for guns. Guns are not needed. We can settle this without guns. Please don't. I told Ebe2 that we would not get the guns but we would not leave until we saw 308's body. The leader did something with the communication device on his belt.

About twenty minutes later, three Ebens showed up inside this building. One of the Ebens identified himself as a doctor and spoke very good English. This doctor had a strange sounding voice, almost like a human's voice. This doctor did not have a high-pitch accent, like Ebe1 and Ebe2. I was very impressed with this doctor. I just wonder where he has been for these past 18 months. We have never saw [seen] him before. This doctor told us that 308s body was not inside the container. The Ebens have done experiments with 308s body because they considered it an honor to have such a specimen to work with. The doctor told us they have used 308s body to create a type of cloned human being. I stopped the doctor at this point. I told the doctor that the body of my teammate was the property of the United States of America, planet Earth. The body did not belong to the Ebens. I did not authorized [sic] *any experiments on the body of 308. I explained that humans consider a body to be religious. Only I could have authorized the use of 308s body for experiments. I demanded to see the body. This doctor explained the body was gone. This doctor said all the blood, [and] body organs were taken out and used to clone other beings. The use of the word beings really scared me and the others. 899 became extremely angry. He called the doctor curse names. I ordered 899 to be quiet. I then told 203 to take 899 out of the building. I realized this matter could really blossom into a major incident. I could not allow that to happen. There were just eleven of us and we realized that if the Ebens wanted to imprison us or kill us, they could do it very easily. But I didn't think the Ebens would resort to such behavior. I was not going to allow this incident to advance into something worse. I realized there wasn't much we could do about what the Ebens have done with 308's body.*

*Ebe2 looked very upset. Ebe2 told me that everyone should be nice, she repeated the word nice many times. Ebe2 did not want this matter to escalate. I kind of felt sorry for Ebe2. She was trying to mediate the matter. 203 suggested we return to our living quarters and have a team meeting. I told the leader that I did not want any further interference between whatever is left of 308s body and experiments. I pointed my finger toward the leader's face. Ebe2 translated, along with the doctor. The doctor, who was extremely straightforward, told me that nothing further would happen with the body, but advised me very little was left of the body. Ebe2 then told me the leader was concerned that we were upset. That we were their guests. That the leader was upset that we were offended. The leader did not wish to upset us and promised that nothing further would happen to the body. I thanked Ebe2 and had her rely [relay] that to the leader. We returned to out [our] huts. Everyone was upset, especially 899. I told each member to calm down. I explained our situation, as if each team member didn't already realize it, that we were only eleven military personnel. We had no way of fighting the Ebens. We did not come 40 lightyears [*sic*] to start a war with the Ebens. A war we could not* *win. We could not even win a simple fistfight with the Ebens. Yes, maybe we could beat them up but what then. We have to realize out [our] situation and act accordingly. I ordered each member to reconsider the situation and to except [accept] the facts about 308s body. I told 633 and 700 to investigate this cloning procedure with the English speaking Eben doctor. Let's get all the facts about what they did with the body and what we can find out about the body and the Ebens' experiments with the body.*

BRAVE NEW WORLD

The trip to Serpo was our traumatic introduction to the world that Aldous Huxley had prophetically anticipated thirty-three years earlier (1932) in his classic novel *Brave New World*. Even twenty years before James Watson, Francis Crick, and Maurice Wilkins had solved the riddle of how human traits are passed on from generation to generation with the discovery of the DNA molecule, Huxley could see where it was all heading. A synopsis of the book in *Masterplots,* edited by Frank N. Magill, says, "human beings were turned out by mass production. The entire process, from the fertilization of the egg to the birth of the baby, was carried out by trained workers and machines. Each fertilized egg was place in solution in a large bottle for scientific development into whatever class in society the human was intended." As early as 1943, Nazi horror doctor Josef Mengele, the so-called "angel of death," was already studying identical twins at Auschwitz to learn how to clone human beings.

By the time the team left Earth in 1965, DNA was well understood by science, and the team scientists should have been aware of it, especially since Watson, Crick, and Wilkins had won the Nobel Prize in Physiology or Medicine in 1962. Also, by that time, there were reports leaking from top secret circles that some of the Grey aliens, who had been abducting humans, were creating a hybrid human-alien race using sophisticated genetic procedures, although the books on this subject by Budd Hopkins and John Mack were yet to be written (see plate 16). Furthermore, it was believed that the Greys themselves were clones.

Time *magazine cover on cloning (February 19, 2001)*

So the concept of human or alien cloning should not have been totally foreign to the doctors and scientists on the team. And yet, they were shocked by the fact that the Ebens considered the body of 308 a great prize for their crossbreeding experiments. This was their rude introduction to the "brave new world" of genetic engineering. But bigger shocks were yet to come.

THE EBEN GENETIC LABORATORY

The Team Commander continues the story in his diary:

> *Ebe2 came to the hut. I told Ebe2 that 633 and 700 were going to examine what was left of 308s body. They would also conduct*

research into the Ebens experiments done on 308s body. Ebe2 look [sic] very concerned. It is sometimes difficult for us, even after the time we have been on this planet, to determine the meaning of facial expressions on [an] Ebens face. Ebe2 replied that she must first obtain approval. [A]pproval was a new word for Ebe2. She was must [sic] reading or learning our language. Maybe she is just picking up our words. I told Ebe2 that she can go get permission but that we were told when we arrived that we did not have any restrictions on where we went. Ebe2 said she would speak with the leader. 633 and 700 gathered test equipment and prepared themselves for the examination of the Eben laboratory. According to our time devices, Ebe2 came back about eighty minutes later. Ebe2 said it was alright for my men to visit their laboratory. I decided that I would also visit it. Myself, 633 and 700 were escorted by Ebe2 to the laboratory facility. We had to be taken by the helitransport device, as we call their helicopters. It took us some time before arriving at this location. According to our compass readings, which are not really a compass reading but we made points of references and according to them . . . we traveled north. The facility was large, by Eben standards. The building looked like a large single story windowless school. We landed on the roof, or maybe just a landing zone on the roof. We were escorted down a walkway or ramp. They don't have ladders on this planet. I think I wrote about this in one of my past entries. . . . They have ramps. We arrived in a room. White walls. We then walked through a hallway into another larger room. We met our English speaking doctor. We saw many other Ebens, all dressed in a bluish-colored one-piece suit. Different from their ordinary suits I spoke of in past entries. The doctor told us that all the experiments done inside this building, he didn't call it a laboratory, just a building, is done to create cloned beings. We were lead [sic] into another room, where there were rolls of containers, looking like glass bathtubs. Inside each bathtub were bodies. I was shocked, as were 700 and 754.

HIDEOUS-LOOKING CREATURES

*Bodies. Strange-looking bodies. Not human bodies, at least not all of them. We started walking down the space between the tubs. We looked into the tubs. These were hideous looking creatures. I asked the doctor what type of creatures were inside these tubs. The doctor told us that these creatures came from other planets. 700 asked the doctor were these creatures dead when they arrived? Or did the Ebens bring them here dead. The doctor said all the creatures were brought to the planet alive. 700 asked if the creatures were kidnaped [*sic*] or brought here against their wishes. The doctor was not sure of the word kidnaped [*sic*]. The doctor seemed puzzled. The doctor asked about the question. 700 said that these creatures were taken from another planet and brought to serpo without permission from them or their leaders of their planet. The doctor said that these beings were brought here for experiments. These creatures are not intelligent beings. Ebe2 then used the word animals. Ok, now I understand. All these creatures are animals from other planets. The doctor didn't seem to understand the word animal. Ebe2 and the doctor exchanged words in Eben and the doctor then said, yes, they are animals. I then asked if there were any intelligent creatures in this building. The doctor said, yes, but that all of them were dead when they arrived on serpo. 700 asked to see these creatures. The doctor corrected 700 by saying beings. Ok, I guess creatures are animals and beings are like humans.*

Let me first write down the description of these creatures inside these tubs. They are not all alike. The first creature I see inside the tub looks like a porcupine. It appears to have a tube placed inside of it. The tube leads to a box underneath the tub. The second creature I see looks like a monster. It has a large head, big deep set eyes, no ears, a mouth, but no teeth. It is about five feet long and has two lower legs but no feet. It has two arms but it doesn't appear to have any elbows. It has hands but no fingers. This creature also has a tube going through it. The next creature looks like nothing I can compare it to. It has blood red skin, two spots in the middle, maybe eyes. No arms or legs. It has a very strange odor. The skin appeared to be blotchy with scales. Maybe like a fish. Maybe it is a fish. The next creature was human like. But the skin was white, not skin white, the color white. The skin was wrinkled. The head was large, with two eyes, two ears, and a mouth. The neck was very small. The head almost looked as if it sat on the lower torso. The chest is thin, with large bone like protrusions. The arms were curled, with hands but no thumbs. The legs are also curled with feet but only three toes. I couldn't look at anymore creatures.

The Egyptian god Anubis—
a human-animal hybrid? (also see plate 23)

I SAW THE DARK SIDE OF THIS CIVILIZATION

*We walked down another hallway, through a room, down a ramp, into another room. We came into a room that looked like a hospital room. There were many beds, or at least sometime [*sic*] of bed, Eben style bed. I described them before. In each bed was a living being, as the doctor called it. The doctor told us that each being was alive and well cared for. 700 asked the doctor if these beings were ill or sick. Ebe2 had to translate that, but the doctor said, no, they are being lived. The three of us [102, 754, and 700] seemed really stunned by that word lived. I asked Ebe2 what the doctor meant. Ebe2 exchanged words with the doctor. Ebe2 then used the word grown. 700 asked the doctor if these were the cloned beings that he mentioned before. The doctor said, yes that each being was being grown, using the same word as Ebe2 just used. 754 asked the doctor if these beings were being grown like a plant. The doctor said yes that is a good comparison. 700 asked the doctor how are they grown. The doctor said that certain parts of other beings were used to grow these beings. The doctor said he could not explain the process in English because he didn't know the words. 700 then asked Ebe2 if she could explain the growing process. Ebe2 said she did not know the English words. Ebe2 said that parts of the blood and other organs are used to mix a substance that is placed inside the bodies of these beings. That was all Ebe2 could explain in English. I told 700 to go back and get 420 and bring them [him] back. While we waited for 420, we looked at these beings. They were breathing. They looked like humans, most of them. Two of the beings on the end looked like humans with dog heads. These beings were not awake. They were either sleeping or drugged. 420 returned. I told 420 to see if he could translate the method used to grow these beings. 420 then spoke with Ebe2. 420 is really good. However long we have been hear [*sic*], some guess about eighteen months earth time, 420 learned the language well. 420 then said that the growing process involved cells taken from other* *beings, grown and mixed with chemicals and then inserted into the bodies of other beings. That was about all 420 could explain. 420 did not know the words Ebe2 used. But the word cells were [was] used. Ebe2 then told me that some things were taken from inside the cells. 700 and 754 then asked if the item[s] take[n] from inside the cells were cell membranes or identification markers for the cells. Ebe2 translated that to the doctor. Both seemed confused and said they could not explain the process because they did not know the English words for it. 700 used the word advance biological extraction of the cell membranes. But neither Ebe2 or the doctor knew anything about that process. I asked 754 if he might understand what they were doing. 754 said that human cells contain smaller substance[s] that can identify the structure with the membranes of the cell. This isn't something that earth technology has advance but 754 has read about it prior to leaving. But 754 doesn't think that earth technology can be used to grown living cells into what the Ebens have done. The Ebens must have found a way to grow cells and to make them into living beings. 700 and 754 said nothing like his [this] is known on Earth. I then asked the doctor if 308s body was used to create a being. The doctor said yes, and showed us the being. I was shocked, as was 700 and 754. This being, with our teammates blood and cells, looked like a large Eben. But the hands and legs were similar to humans. How could they have grown this being so quick. Obviously, this is well above our intelligence. I saw all I wanted to see. I told the doctor that we would like to leave. Ebe2 saw that I was upset and touched my hand. Instantly, I felt concern. Ebe2 was really concerned about what I saw. Ebe2 said we leave. We traveled outside this building, a building that I did not wish to see again. I saw the dark side of this civilization. The Ebens are not the humane civilization we thought they were. But I must say they didn't hide anything. The doctor spoke straight to us. Just like the Ebens. They don't know how to lie. Seeing what we saw will change our impression of the Ebens as long as we stay on this planet.*

The two doctors, 700 and 754, were beginning to understand what was going on when they asked if the items taken from inside the cells were cell membranes or identification markers for the cells. And then when 754 said that "human cells contain smaller substance[s] that can identify the structure with the membranes of the cell," he was describing DNA without being aware of it. The commander goes on to say, "This isn't something that earth technology has advance[d to], but 754 has read about it prior to leaving. But 754 doesn't think that Earth technology can be used to grow living cells into what the Ebens have done . . . 700 and 754 said nothing like this is known on Earth." The doctors basically understood what was happening, but DNA biotechnology on Earth was not yet being used to duplicate living beings or grow organs. They must have obtained a quick education about this when they saw the hybrid creature that had been created using 308's DNA.

Evidently, the Ebens have the same capabilities as the Greys. Their ongoing alien genetic engineering program has been proceeding since the 1950s in underground laboratories, such as at Dulce in New Mexico. Various whistle-blowers have reported seeing the genetically engineered "freaks" in these locations. We have, no doubt, learned a lot from these alien experiments, and now, with our well-developed DNA knowledge, we basically understand, at least theoretically, what the Ebens and Greys are doing. Whether or not we have attempted to duplicate any of their monstrous experiments is, of course, unknown. That may be one reason for the continuing secrecy.

13

A POLICE STATE

In the following entries in the Team Commander's diary, we find that there was no official program to exchange scientific information at the highest levels, but the team was reduced to teaching technical information about Earth to some uncomprehending students. This is amazing. It would have been expected that the team scientists would have been invited to discussions with the top Eben university and organizational technocrats. Evidently, the Ebens had obtained all the information about our science and technology that they wanted by covert scrutiny of our university and corporate research facilities. We really didn't know very much about the extent of their surveillance activities prior to the Roswell crash. Apparently, they had concluded that we were still in the scientific dark ages, and they weren't interested in anything that we had to teach them.

HOW DO WE EXCHANGE SCIENCE?

In this diary entry, the Commander wrestles with teaching Earth science to some Eben students:

> *It was difficult to speak science to the Ebens. How do we explain Einstein? How do they explain their Einstein? We had a difficult time relating our science to them. However, they seemed to understand our physics and chemistry quicker than we can. We managed to observe some strange things about their technology. First, we*

*took apart one of the locators they placed on our belts. It wasn't easy. There were no screws or bolts holding it together. We had to break the thing. The electronics inside was nothing we have ever observed. There were no transistors, tubes, rectifiers, coils, or other electronic components like our technology. This thing just had wires and some bulges in the wires at certain points. There was two components, which none of use [*sic*] had ever seen. We could not use our frequency counter to determine the frequency it transmitted and received. It was out of our range. 633 and 661 used some other equipment to analyze the thing but couldn't understand it. We asked the Eben Scientist, who we called Ebe4. The problem was the translation Ebe2 had to translate because Ebe 4 didn't speak English. A [lot] was lost in translation even though Ebe2 does a pretty good job with English.*

We showed Ebe 4 one of our portable radios. The Motorola FM radio was pretty complex to us. It was new and contained four channels. 661 took apart the radio in front of Ebe 4, explained the parts and the different crystals we use for frequency. Ebe 4

1960s-era military-style Motorola FM field radio

couldn't understand it. He seemed as lost with our radio as we did with theirs. Ebe2 told us that Ebe 4 couldn't understand the radio or how it worked. So that was our dilemma. How do we exchange science Each of our civilizations must learn from the other. So we decided to begin a school. Our first days were pretty tough. We started with simple things, which we thought would be similar to what they would know. We chose light. 661, who did some teaching before, started with wavelengths. 661 started with nonvisible light and the different angstroms. Then 661 showed them the spectrum of light [see plate 18]. 661 showed them cosmic rays and how we measured them. Then he went to gamma rays, x-rays and ultraviolet light. 661 explained that light was what we called electromagnetic waves. Over a period of eben days, 661 explained everything he knew about light, frequencies and a description of the frequency bands.

AN EBEN EINSTEIN

During this time, several other Ebens came in and listened. Ebe2 was extremely tasked with translating this. Ebe2 had a difficult time explaining everything 661 said because she didn't know the Eben words for everything. But she did an outstanding job of describing what 661 was saying. I don't think Ebe 4 got everything 661 said but it didn't take long for Ebe 4 to realize what 661 was describing. 661 then showed Ebe 4 a repair manual for one [piece] of our test equipment. Since everything, or almost everything we brought was military, it was a military manual. The manual contained schematics of the circuits. Ebe 4 was completely lost. But eventually figured out what 661 was showing him was the inside of the test equipment. 661 then started showing the basics of electricity. Ohms laws, different formulas to figure voltage and amperes. Ebe 4 was confused, to say the least. But one of the other Ebens who had come in to listen grasped the idea quickly.*

*See appendix 2 for a list of all the test equipment brought by the team.

We called this Eben Ebe 5 or the Einstein. This Eben was smart, exceptionally smart. After three years, we finally meet an Eben who can grasped [sic] our science. The only problem is he couldn't speak English. But he asked questions, which Ebe 4 didn't really do. Although it took several lessons to teach Ebe 5 what each letter meant in a formula, Ebe 5 finally realized what we were saying. This Eben must have an IQ of 300. Ebe5 actually solved some simple problems set up by 661. Basic electricity, solving for the amount of resistance in a circuit, simple things like that. It was a remarkable scene. Ebe 5 became our ace student. We couldn't get rid of the guy. He followed us around and asked questions through Ebe2. When Ebe2 wasn't available, he would simple [sic] point and shrug his shoulders. We'd speak to him in English or have 420 or 475 translate to him. But only 420 could understand most of what Ebe 5 was saying.

EFFICIENT MANUFACTURING

This come [sic] to another interesting point. Ebe 5 looked a little different than the rest of the Ebens. During the last several years, we have noticed some of the Ebens, especially the ones who live up North, look different. There [sic] heads seem to be bigger. They have a more weathered looking face. Ebe 5 is from the north. He lives in the second village to our north. The distance is approximately 5 kilometers. . . . I drew the map in entry 4432, showing all the villages in the north. I'm sure they have more villages farther away but we haven't visited them yet. Ebe 5 also does not have a mate. This is strange but not totally uncommon. We have found several Ebens without mates. We haven't delved into the personal life of Ebe 5 but 518 wants to. I explained in one entry about the Eben technology when it comes to screws, bolts, etc. They don't have any. Everything they make is sealed with some sort of solder or melting method. When we visited their manufacturing plant, we were amazed at how efficient they were with manufacturing furniture, their helicoptersor flying devices. We still *haven't seen their main spacecraft manufacturing plant.* It must be in the far west and south. I'm sure we will visit the place someday. We still have seven years, or at least seven earth years left. As I mentioned before, we have totally lost track of Earth time. We gave that up many of our years ago. We have gauged our time on Eben time, which is extremely complex, as I mentioned before.*

In the following entry in the Commander's diary, we begin to appreciate how an alien civilization could have achieved a technology so advanced that they can travel through time to distant stars, create something as wonderful as the Yellow Book, and can clone complex organisms, and yet is not aware of the elementary principles of electricity that would appear to be the essential prerequisites of such technology. As with the Eben doctors who seemed to have a monopoly on advanced bioscience, it is probable that the scientists who understand space travel and navigation are also a small elite clique. In other words, the Ebens apparently reserve higher science and technology for a limited few, and do not share this knowledge with the masses. They may even have initiation rites for these selected individuals. This is reminiscent of ancient Egyptian civilization, where only the high priests were taught hieroglyphics, the language of the hierophants, and of modern America, where advanced science is contained within the classified world of "black projects," available only to those who have passed complex background checks, and so have been admitted to the ranks of the scientific "elite." This seems to be a method of social control and manipulation not limited to Earth. The fact that the sophisticated technology based on advanced physics and chemistry does not trickle down to the masses is exemplified by this statement by Anonymous, "The Ebens had no forms of refrigeration, except in industry." In other words, the Eben civilization is basically a military-industrial oligarchy. The population is divided into two groups—the

*A pattern of Eben secrecy seems to be emerging. They may have had a security reason for keeping the team away from this facility.

elite controllers who jealously guard science and technology, and the masses, who are manipulated and kept in the dark, but to whom are doled out, in a highly controlled manner, some of the fruits of the advanced technology. This became confirmed when the team learned more about the Eben government.

The conclusion that the Eben masses are strictly controlled is supported by the following remarks by Anonymous:

> Athough [*sic*] the Eben civilization had no televisions, radios, etc., each Eben had a small device belted to their waists. This device gave orders to perform a particular task, news of pending events, etc. The device displayed a screen, similar to a television screen but in a 3D style format. Our team brought back one of these devices. (I think today, we could compare it to a palm pilot.) [This parenthetical remark by Anonymous was made in 2006. Today, it would be comparable to a smartphone.]

Anonymous continues:

> The visitors [the Ebens, since they originally visited Earth] were extremely disciplined in their daily lives. Every visitor worked on a schedule, which was not by a clock, but by the movement of their sun. Each little community had a large tower, which filtered the sun through. When the sun was at a particular point on the tower, it meant the visitors had to do a particular thing.

Since they had no planetwide communications access, the Ebens amused themselves with locally based, simpleminded games and distractions. Anonymous says:

> There were no televisions, radio stations, or anything like that. The Ebens played a game, something like soccer, but with a larger ball. The object was to kick the ball down a field into a goal. The game had very strange rules and [was] played for long periods of time. They also had another game, mostly played by the children, that consisted of making formations with groups of Ebens. They seemed to really enjoy the game,

> but our team found little understanding of the game . . . Every Eben was issued what they needed. No stores, malls, or shopping locations. There were central distribution centers where Ebens went to obtain items of need. All Ebens worked in some capacity.

We learn more about Eben government from these statements by Anonymous:

> The Ebens had no single ruler. There was a Council of Governors, which the team named. This group controlled every single action on the planet. The members of the council seemed to have been around for a long time . . . There were leaders, but no real form of government. There was virtually no crime seen by the team. They had an army, which also acted as the police force. But no guns or weapons of any type were seen by our team. There were regular meetings within each small community. There was one large community, which acted as the central point of the civilization. All the industry was at this one large community. There was no money.

Evidently, the actions of every Eben were controlled by orders or directions from a central authority via the device worn on the belt, and by the movements of the sun. The central authority is apparently an agency of the Council of Governors, so named by the team. This makes the Eben populace almost robotic. It would even be appropriate to use the term *enslaved.* They certainly aren't catered to. All their basic needs are met, but they are not really free. They live very uniformly Spartan lives. Anonymous says:

> The individual Eben family lived a simple life. Their homes were constructed of clay, some type of material similar to wood, and some metal. The houses all looked the same. They appeared to be something from the Southwest, looking like adobe. The interior of the house consisted of four rooms. One sleeping room where all Ebens slept in the same room on mats, a food preparation room (kitchen), a family room (the largest in the house), and a small waste room.

Not only did they all act and live alike, they also looked alike. Anonymous says:

> Everyone looked the same initially to our team. But after some time, the team members learned to identify different Ebens by their voices.*

The picture of Eben civilization that emerges from this information is one that resembles the future society in *The Time Machine* by H. G. Wells, where the Eloi populace was indoctrinated by the Morlocks to believe that they were free, but were actually hypnotically controlled to go through robotic activities and to periodically become food for their underground masters. This is also reminiscent of Nazi Germany where the people were not permitted to listen to foreign radio broadcasts so that they wouldn't be able to learn what life was like outside of Germany, and could thus be completely controlled by propaganda. At the same time, their elite SS masters were developing space-age technology and an advanced society in Antarctica, which the German populace knew nothing about.

A STRUCTURED CIVILIZATION

Anonymous answered some of the questions that were sent in to the website after it was launched in November 2005. The following information was given in answer to a question about the Eben population.

> Why does the Eben population number only about 650,000? The Ebens have a very stable, structured civilization. Each male has a mate. They are allowed to reproduce (in somewhat the same sexual way we do), but are limited to only a specific number of children. Our team never saw a family with more than two children. The Eben civilization was so

*Since we have already learned that the Ebens had perfected cloning technology, it is not out of the question that many of them were cloned. We know that the Greys relied on cloning to increase their population because they had lost the ability to reproduce. Perhaps the Ebens were starting to experience a similar problem, and used cloning to supplement their numbers, while still controlling normal reproduction.

> structured that they planned the birth of each and every child, spacing them apart to allow the proper social grouping of the civilization. Eben children matured at a super rate, compared to Earth children. Our team watched live births, attended to by an Eben doctor, and then watched the development of the child over a period of time. . . . They matured at an alarming rate.
>
> The Ebens had scientists, doctors, and technicians. There was one educational facility on the planet. If one was chosen, you attended the facility and learned the job one was best qualified and suited for. Although it was extremely difficult to judge or measure, the team estimated each Eben's IQ to be 165.

This information further substantiates the conclusion that the Council of Governors exercised complete authority and control over every aspect of Eben life. There appeared to be no appeal from their decisions. It was absolutely dictatorial. And yet, the team estimated that the most of the Ebens were extremely intelligent. The novel concept of individual liberty had not yet come to Serpo. They were all expected to work, and the fruits of their labor accrued primarily to the ruling class. Deprived of any leisure, they were not able to produce artistic or creative works to entertain or enlighten each other. The world they inhabited was grim and gray. And yet, we can dimly perceive the seeds of rebellion in Ebe5, the "Eben Einstein," who followed the team around. Apparently, he began to appreciate the concept of liberty in the actions of the team members and wanted to know more about their principles of freedom.

A "1984" SOCIETY

This excerpt from a later diary entry by the Commander reinforces the view that Eben society was strictly controlled:

> *At one of our team meetings recently, 203 and myself decided to give up the military greetings, saluting that each member gave the first time they saw us. I decided we maintain our military bearing*

and manners but we will give up the saluting. Each team member agreed. I have no problem with that. The Ebens just starred [sic] *at us when we did that.*

But they also have their greetings. Ebens exchange greetings depending on the time of day. They hug on certain times, touch fingers on other times and bow at other times. We still haven't figure [sic] *out why they do that. Ebe2 just explained it was a formal greeting method.*

Ebens live a very strict life. They keep a regimented lifestyle. We have seen some variances, but only by a few. The military keeps everyone in line. The military acts as a police force, as I mentioned before. They do not carry any weapons but they do have different uniforms and every Eben respects that uniform. The military are patrolling all the time. They walk in pairs and seem to be extremely friendly, but can be very strict. We saw two Ebens walking across a field. Two military members quickly approached the two Ebens and pointed to a building. The two Ebens walked to that building with the military members. The military members were yelling something at them. At the time, neither 420 or 475 were available for translation but I figured those two Ebens violated some custom or law. We have been warned by the military when we approach something we are not suppose [sic] *to approach. The military are very respectful when they deal with us but they don't allow us to violate any of their customs or laws without warning us. When we first killed one of the sand snakes, we had six military members at the scene in no time. It took a great deal of diplomacy to deal with that situation. But the military never touched us nor threatened us.*

The Ebens have adjusted to us, just as we have adjusted to them. We carry on our mission and they allow us to do just about anything. The one forbidden act is not to enter a private living house. We did that once and was politely escorted out by the military. There seems to be more military than are actually needed. They do have weapons, as I mentioned before. We rarely see any military member with those weapons. But we have seen them during the alert that occurred some time ago. Ebe2 came into our living compound, just after one of our rest periods. Ebe2 was excited and told us to stay inside and not to exit our living quarters. We asked why and Ebe2 said that an unknown spacecraft had entered their planet's orbit. But Ebe2 assured us that the military would take care of the problem. We naturally went into our own alert. We issued our weapons and stood by to guard our living area. We violated her instructions and went outside. We watched the skies and saw a lot of air traffic. We then saw all the military members with weapons and something that looked like field packs. They were in full fighting gear, as 899 called it. The alert didn't last very long and Ebe2 came back, looked at us a little curious and then told us everything was alright and the alert was over. We asked her if the unknown spacecraft was identified. She said it wasn't a spacecraft, but just a natural piece of space debris and left it at that. We didn't believe her, but we had no way of knowing any difference. We returned to our normal routines.

With this new information from the Team Commander, we are now justified in referring to Serpo society as a "police state." When he says, "The military keeps everyone in line . . . The military are patrolling all the time," we can reasonably conclude that the police state designation is appropriate. It's unfortunate that we are not told how the ruling class lives, but we can presume that their lifestyle is far more commodious, perhaps even luxurious. Since the doctor in the biomedical building spoke perfect English, and he was clearly a member of the upper class, we now know that they are very well educated, especially about Earth matters. But, from what we have learned so far, our team was not permitted to meet them.

Evidently, tyranny and enslavement exist in many worlds in our Milky Way. We can see clearly that this would have been our fate too if the fascists had won World War II. Now, as we begin to venture out into the cosmos, we must take it upon ourselves to embark on a mission to bring the blessings of liberty to the oppressed peoples throughout our galaxy. It's the right thing to do. Thomas Jefferson would have

The Star Trek *team helping to liberate the galaxy from oppression*

encouraged us to become the liberators of galactic societies subjugated by tyrants. Perhaps if, in the near future, William Shatner were elected president of the United States, we might be able to join our space brothers in the Galactic Federation in their quest to restore, protect, and advance freedom throughout the galaxy. Certainly, we would be a highly valued asset with Captain Kirk at the helm.

14

FEAST, FUN, AND DEATH

In answer to a question submitted to the website, Anonymous says:

> As to the Eben culture: They had a form of musical entertainment. The music sounded like tonal rhythms. They also listened to a type of chanting. The Ebens were dancers. They celebrated certain work periods with a ritual dance. The Ebens would form a circle and dance around, listening to the chanting type of music. The music was played on bells and drums, or something similar to them.

From the Commander, we learn more about how Ebens have fun:

> *We had a feast today. What a feast. We used our last C-Rations but the Ebens didn't really care for our food. We did kill a beast. As I mentioned before, the Ebens allowed us to kill the beasts for meat. The meat isn't really bad, 899 says it tastes like Bear, which I never ate. But Ebens look at us very strange [*sic*] when we eat meat. Strange, they can clone creatures and other species of humans, but they can't eat meat. How strange they are. But they allow us to do just about anything we want, and eating meat is something we need for the protein. We used the last of our salt and pepper, which does make eating their food more of a challenge. The Ebens don't have anything similar to salt and pepper. They do have an herb, as we call it, something like Oregano which they use. It has a tart taste but we have developed a taste for*

it. The feast was great. We participated in the dances, which the Ebens really like. They get great pleasure dancing and playing their strange games. I described the games before, but [at] this feast, we saw something different. The game was played like chess, but with Ebens standing on a large square area of the compound. The squares were divided into twenty-four sets. Each set had another two spots. Just how or why the Ebens moved was a mystery to us. One of the Ebens would say a word and then another Eben would move. It appeared that it was a team game. Six Ebens on each side. We couldn't figure it out but at the end, the Ebens danced with each other, signifying a victory, we think. It was a fun day.

BALL GAMES AND SEX

And more from Anonymous:

Our team brought along softball equipment for sporting activity. The Ebens would watch the game and laugh out loud. (The Eben laugh sounded like a high-pitched yell.) Eventually, the Ebens started playing the game, but never got used to catching the ball before it hit the ground. Our Team also played touch football. Again, the Ebens watched the game intensely and then played it themselves. But again, like softball, the Ebens never figured out they had to catch the football before it hit the ground!

Although our Team Members honored the privacy of the Ebens, our Team was allowed to witness births. Our team, snooping aroud [*sic*], was able to capture the sexual activity of the Ebens. The males and females had similar sexual organs and performed intercourse. The frequency of sexual activity was not recorded as being as often as in our society. It was believed that they performed the act for pleasure and reproduction.

The Team Commander continues his narration about the feast day:

My team played softball with some hard nosed Ebens, who have learned the game. Well for the most part. They still haven't figured out they have to catch the ball before it hits the ground. But they had fun. We have found some extremely gifted athletics [athletes] among the Ebens. Then again, we found some who had no athletic ability. Just like humans. Our softball game ended when the rains came. We ended up inside the community building. We finished our meals and went to our living area. As we do each day, we have our end of day briefing. We check each other for psychological condition and medical health. Our day ends and we start our eight hours of rest.

WE HAVE GIVEN UP EARTH TIME

*Ebens have a different period of time, as I mentioned before. They rest about four hours, for every ten hours' working time. But we must consider that their hour is longer and that their days are longer. So we stopped using our time and used the Eben time. It is difficult to understand, but this is only a diary. Once I return to Earth, I can explain the time difference and how we had to use their time instead of ours. I keep writing about the time in every diary entry, but it is important to note that even though we have been hear [*sic*] for about three years, Earth time, we have given up Earth time and utilized the Eben time. We tried to use their two suns as a counting system, but that didn't work. We then tried to use our own watches, but that didn't work. So we gave up our timepieces and just use the Ebens' time tower. Each village has one and it is easy to understand the symbols. Each symbols [*sic*] means a certain time and certain work period.*

The time situation was complex because it related to the way Serpo revolves around Zeta Reticuli I, while influenced by Zeta Reticuli II. Here is what Anonymous says about the time problem:

Our scientists had the same questions, as posed by your audience. Our scientists questioned our team members and the information they gathered. Our scientists could not understand how the orbit of . . . Serpo could revolve around the two suns at the distance measured. In the end, our scientists found that some things relating to that particular system was different in physics compared to our system. There were some questions about how our team measured the orbit and other calculations based on the lack of a stable time base. For some reason—and I don't think this was ever determined—our time instruments did not work on Serpo.

Now, considering this, you can understand the difficult job our team members had making calculations without time. They had to come up with an alternate method to measure speeds, orbits, etc. Challenge: Try solving a problem in physics without being able to measure time on Earth! So you see, our team did the best they could with the instruments they had and the hardships they developed attempting [to make] scientific calculations. It is difficult for any Earth-based scientist to understand the different physics in other solar systems or on other planets. One of the questions sent me involved Kepler's Law of Planetary Motion. Our team had that information. We had some of the best military scientists on the team. But if you consider Kepler's Law, it requires time and our team could only measure time in the conventional way. It was determined that Kepler's Laws did not apply to that solar system. CONCLUSION: One of the things our Earth-based scientists learned was not to apply Earth's laws of physics in a universal way.

Anonymous says further on this subject:

Regarding Time: The Team Members brought several timepieces, e.g., wristwatches, nonbattery style, as it stated in the debriefing data. The time pieces worked, but they had no reference to time since the Eben days were longer, the dusk and dawn periods were longer and they had no calenders to reference. They did use the timepieces to calculate movement; for example, timing the movement of the Eben two suns. They also calculated the time between work and rest periods. But, after awhile, the team discarded their timepieces and used the Eben's

Johannes Kepler (1571–1630) (also see plate 24)

> measurement of time periods. The team became confused with the calenders [*sic*] they brought—a ten-year calender [*sic*].
>
> After 24 months, the team lost track of time, as to the calendar [*sic*] since they could not properly calculate days compared to Earth days. They set up one large clock to the earth time when they left. However, this was a battery-controlled clock and when the battery died, the clock stopped and they forgot to change the battery in time. Consequently, they lost the earth time. The team brought a large quantity of batteries, but they ran out after about five years. The Ebens had no comparable item like batteries.

Noted Cornell astronomer Carl Sagan was consulted about the planetary motion of Serpo that contributed to the timing difficulties. About this consultation, Anonymous says:

> One of the principal home-based scientists (astronomer) contracted to assist us was Dr. Carl Edward Sagan. Initially, he was the biggest skeptic of the group. But as information was slowly analyzed, Dr. Sagan came back to the middle. I can't say he fully accepted every single piece of data, but he did agree on the final report.

E-mail moderator Victor Martinez supplied the following "FAST FACTS" about Carl Sagan's involvement in the program, on the website:

Born in Brooklyn, NY on 11-9-34 and died in Seattle, WA on 12-20-96 of bone marrow cancer. He was an American astronomer, educator, and planetary scientist and was the director of the Laboratory for Planetary Studies at Cornell University.

CONNECTING THE DOTS: Project Serpo's final report was written in 1980 with Dr. Sagan having been brought in halfway through the project. It is believed that he wrote his 1985 best seller, CONTACT, based on his insider knowledge of the most secret project in human history: a human-alien exchange program of which he signed off on its final report! Years later, his book was made into the 1997 movie CONTACT starring Jodie Foster.

Sagan became well known as a result of his public debate in 1969 with Dr. J. Allen Hynek, sponsored by the American Association for the Advancement of Science (AAAS), about whether or not UFO investigation should be considered serious science. Sagan argued that it was pseudo-science, and he was proclaimed the winner. He was brought in as a consultant on the Serpo project about a year later, when he then had to, of course, reconsider his skepticism. Sagan also locked horns on the same issue with well-known scientist–UFO investigator Stanton Friedman, his classmate in physics at the University of Chicago in the

Carl Sagan, famous Cornell astronomer widely believed to be skeptical about UFOs

1950s. But, even long after his involvement with the Serpo program in 1980, when he wrote a section of the final report, he was still able to say, in his bestseller *Cosmos,* published in 1985: "We maintain that there is no credible evidence for the earth being visited, now or ever."

We now know posthumously that Sagan really wasn't two-faced about the extraterrestrial question. He had to be skeptical about UFOs in public so as not to jeopardize his position in the astronomy department at Cornell, where he could not afford to appear to be unprofessional. Cornell relied heavily on government funding for its astronomy research, especially from NASA, and this could have been terminated had his real beliefs been revealed. In reality, Sagan was a member of the Council on Foreign Relations, and it is believed that he may even have been a member of MJ-12! His true interest in extraterrestrials was revealed in his blockbuster book, *Contact,* later made into a hit movie in 1997, starring Jodie Foster and Matthew McConaughey. Sagan was a man of many talents. He won the Pulitzer Prize for nonfiction in 1978 for his book *The*

Contact *movie poster*

Dragons of Eden. He has been immortalized in downtown Ithaca, New York, the home of Cornell, with the Sagan Memorial Walk of Planets. Ironically, in the end, it was he, more than anyone else, who raised the public consciousness all over the world about the possibility of life on the "billions and billions" of other worlds. But he did it his way. In terms of public perception, he had to walk a narrow path. But he navigated it successfully, and his continued fame is well deserved.

The problem of Serpo's planetary motion as related to the measurement of time generated a flood of commentary to the website. See appendix 4 for a summation and encapsulation of this commentary, with a layman's attempt to comprehend and bring some order to the discussion.

THE DEATHS OF 899 AND 754

The Team Commander's diary entry continues:

> *Ebe2 came by after the feast. Ebe2 was concerned with 754. As I mentioned in one past entry, 754 became sick. But he has recovered. We don't know what he suffered but 700 treated him with penicillin, which worked. We all have had some sort of sickness, since we've been here except 899, that guy is a solid rock. He hasn't been sick, not even a cold. 706 [700] and 754 are keeping details [*sic*] records of each team member and their medical and physical condition. We have tried to keep a steady physical fitness program since we arrived. Sometime we follow it and sometime we don't. But everyone is in pretty good shape, at least physically. Mentally, that might be another story. Some team members miss Earth, as I do. But no team member has broke [*sic*] down or has needed any type of psychological help from 700 or 754. Our screening process was great. Keeping busy is our medicine. We keep extremely busy, exploring and doing our mission goals.*

This diary entry by the commander was made about three Earth years into the visit. Shortly thereafter, two of the team members he mentions died. The security man, 899, was the first to die on Serpo. Evidently, his death occurred suddenly sometime after this diary entry was made, wherein, ironically, the Commander had said that he was "a solid rock," and "He hasn't been sick, not even a cold." Anonymous gives a complete narration about the death of 899, and the attempt to revive him by the Eben doctors.

> When our first team member died in an accident, it was hard to communicate with the Ebens. The member died instantly, therefore, no medical care was provided. Our two doctors examined the member's body and determined the injuries were consistent with an accidental fall. Initially, the Ebens never interfered with our care or offered to provide any of their medical care. However, once the Ebens—a very benevolent and caring people—saw our team members crying, the Ebens stepped in and offered to attempt some sort of medical care. Although our doctors felt the team member was medically dead, they allowed the Ebens to try their own medical care. Most of this was either through sign language or speaking to the Travelers who could understand some English.*
>
> The Ebens transported the team member's body to a remote area of the largest community. They took the body into a large building, apparently their hospital or medical center. The Ebens used a large examination table to view the body. The Ebens ran a large bluish-green light beam over the body. The Ebens watched a display that appeared on a large screen that looked like a television screen. The readouts were in the Eben written language and thus our team could not understand it. However, there was a graphic display, similar to a heat [heart] beat graph. The solid line was not wavering. Our doctors understood that meant the same thing that their equipment measured: the heart was not beating. The Ebens administered some liquid through a needle. This was done several times. Eventually, the heart started beating.
>
> But our doctors knew the internal organs of the body were

*The Travelers were Ebens who had demonstrated some facility with English.

> damaged, but couldn't fully explain that to the Ebens. The Ebens finally made a sign, placing both their hands to the chest and bowing their heads. Our team members knew that meant the body was dead and nothing could be done. The Ebens showed affection to our team. During the last work period, the Ebens had a ceremony for the dead Team Member, the same ceremony used when an Eben died. Our team held their own service, attended by the Ebens. The Ebens were extremely curious about our religious service. One Team Member, who was acting as a minister, performed a death service. Our Team was eternally greatful for the Eben's caring attitude for our dead friend.

The second death also occurred sometime shortly after the above diary entry about the feast was made by the Commander. One of the doctors died of pneumonia. We know he was alive at the time that diary entry was recorded because both doctors are mentioned. It would be fair to conclude that it was 754 who died, since we learn in that entry that he had been sick and treated with penicillin by 700, the other doctor. Interestingly, it was Ebe2 who first noticed, after the feast, that 754 was sick, and she was concerned about him.

EBENS DID DIE

It is not surprising to find widespread and standardized religious observance in a police state, as in Mussolini's Italy. It is actually expected, because it is usually encouraged by the ruling authority. After all, it is easier to keep citizens in line when they believe that the social rules and regulations come from some sort of supreme being. The fact that all the Ebens were required to attend worship services every day at a designated time leads us to believe that this was another form of control by the ruling class, especially since it was uniform throughout the planet. Where there is freedom of belief, divergent opinions would be expected.

From the following remarks by Anonymous, we learn about Eben religious customs. All of this information was taken from the Debriefing Document.

Ebens did die. Our team members saw deaths—some from accidents and some from natural causes. The Ebens buried the bodies, similar to our method. Our team saw two air accidents involving their intraplanet flying vehicle. The Ebens worshipped a Supreme Being. It appeared to be some sort of diety relating to the Universe. They conducted daily services, normally at the end of the first work period. They had a building or church they entered to worship.

THE EBEN WAKE

Our team witnessed an aircraft accident that killed four Ebens. The Ebens performed a form of ritual at the crash site. The Ebens transported the bodies to a medical facility and examined the bodies. Our team members were always allowed to accompany the Ebens, except during rest period, when the Ebens closed their doors for privacy.

Our team members saw the sorrow in the eyes of the Ebens during the death of their own. Later, after the last work period of the day, the Ebens had a "funeral," at least that is what our team concluded it was. The Eben bodies were wrapped in a white cloth. Several types of liquids were poured over the bodies. Large numbers of Ebens would stand in a circle, chanting. The sounds became almost nauseating to our team members. The ceremony lasted for a long time. Finally, the bodies were placed in metal containers and buried in a remote location away from the communities. After the burial, the Ebens had a feast. Large tables of food were brought out and everyone ate, danced, and played games. This occurred at every Eben death witnessed by our team.

From the foregoing comments by Anonymous, we learn that the concept that death is a joyous release of the soul from the cares of physical existence may well be a galaxy-wide belief.

STAR WARS

Anonymous tells us about their Great War. Perhaps George Lucas knew about this piece of Eben history when he wrote *Star Wars* (see plate 19):

> . . . many hundreds of thousands of Ebens died in the Great War . . . The Ebens fought a battle with an enemy for a period of time. Our team members estimated the war lasted about 100 years, but, again, that is our time. The war was fought using particle beam weapons, developed by both civilizations. The Ebens eventually were able to destroy the enemy planet, killing the remaining enemy forces. The Ebens did warn us that several other alien races within our galaxy were hostile. The Ebens stay away from those races. The debriefing document never stated the name of the enemy, probably because they no longer existed.

This information helps us to understand why the Eben military was so strong and dominant. Evidently, the aftermath of their Great War left the populace traumatized, and very willing to accept continuing military authority. For more about particle-beam weapons, see chapter 17.

15
EXPLORATION

In this chapter, Anonymous gives details about the history and the physical environment and characteristics of Serpo. Information in brackets was added by Victor Martinez, the e-mail moderator. See appendix 3 for complete statistics about Serpo.

Anonymous says:

> Serpo was estimated to be about three billion years old. The two suns were about five billion years old, but only by estimation. The Eben civilization was estimated to be about ten thousand years old. They evolved from another planet, not on Serpo. The original home planet of the Ebens was threatened with extreme volcanic activity. The Ebens had to relocate to Serpo in order to protect their civilization. This occurred some five thousand years ago . . .
>
> There was a period of darkness, but not total darkness.* The Eben planet is located within a solar system of the Zeta Reticular [Reticuli] Star System [two 5th-magnitude yellow double stars, similar to our sun, located near the Large Magellanic Cloud]. The planet had two suns but their angles were small and allowed some darkness on the planet depending on one's location.
>
> The planet was tilted, which allowed the northern part of the planet to be cooler. The planet was a little less than Earth's size. The atmosphere

---

*A comment sent in by Paul McGovern confirms this. He says, "There was never complete darkness on the visitors' planet. It got dim but not dark."

was similar to Earths and contained the elements of CHON [carbon, hydrogen, oxygen, nitrogen]. Zeta Recticular [Reticuli] is approximately thirty-seven light-years from us . . .

There were about a hundred different villages or living locations for the Ebens. The Ebens only used a small portion of their planet. They did mine minerals in remote areas of the planet and had a large industrial plant in the southern portion of the planet near a body of water. Our team determined this plant had some sort of hydroelectrical operation.

Paul McGovern adds, in his comment, "There were small communities throughout the planet. There were underground rivers, which fed into open valleys."

THEY COULD TRAVEL AS THEY WISHED

Regarding the team's adaptation to the conditions on Serpo, Anonymous says:

Once the team arrived on the Eben planet, it took them several months to adjust to the atmosphere. During the adjustment period, they suffered headaches, dizziness, and disorientation . . . The bright suns of the Eben planet also presented problems. Although they had sunglasses, they still suffered from the bright sunlight and the danger of sun exposure. The radiation levels [level] of the planet was a little higher than that of Earth. They were careful to cover their bodies at all times . . .

Our team eventually relocated to the north in order to stay cool. The ground transportation used by our team was similar to a helicopter. The power system was a sealed energy device that provided electrical power and lift for the craft. It was very easy to fly and our pilots learned the system within days. The Ebens did have vehicles, which floated above the ground and did not have any tires or wheels [see plate 20].

Paul McGovern adds the following information in his comment: "The team was never isolated or restricted by the visitors. They could travel as they wished and see whatever they wanted to see. After about six years, the team moved to a northern portion of the visitors' planet, where the temperature was cooler and which contained ample vegetation . . . The visitors built a small community for the team . . . The exchange team had to endure extreme hardship adjusting to the environment of the visitors' planet . . . The heat was extreme and took many years to adjust."

Anonymous describes Eben technology:

> The Ebens developed a different type of electrical and propulsion system. It was unknown to our team and I don't think we ever really understood it. They were able to tap into a vacuum and bring back an enormous amount of energy from that vacuum [see chapter 17]. Our team's living quarters, which consisted of several small buildings, contained electricity powered by a small box. This small box supplied all the power our team needed. Ironically, the electrical equipment our team brought on the trip worked using their power source only . . . The Ebens energy device was analyzed over and over again by our team. Since our team did not have access to scientific microscopes or other measuring equipment, we could not understand the function of the energy device. But, regardless of the electrical demand, the Eben energy device provided the proper current and wattage. Out team surmised the device had some sort of regulator that sensed the required current/wattage and then supplied that specific amount. (Note: Our team members brought back two energy devices for analysis.)

Los Alamos actually had one of these energy devices in their possession as early as 1947. It was retrieved from the second crashed Roswell disc, but no one had any idea what it was. It wasn't until 1970 that they finally realized that it was some sort of power source, but they could not understand how it worked. Then, when the Serpo team brought two of the devices back to Earth in 1978, experimentation was begun in earnest to understand how it functioned, and attempts were made to duplicate the technology. It was given the name Crystal Rectangle, or CR for short. A small dot was visible in the CR that moved whenever an electrical demand was made on the device. After years of experimentation and research, still ongoing, it was determined that the dot was a perfectly rounded particle of charged antimatter. For a complete history of the experiments involving the CR and the duplication efforts, see appendix 5.

Anonymous continues:

> They [the team] also took electric razors, coffeepots, electric heaters, a DIM (no explanation as to what this was), an electric IBM typewriter, a scientific calculator, slide rules (both conventional and scientific), Base Data Collection Recorder (BDCR), three different sized telescopes, tangents, both conventional and electrical. The list goes on and on. But they took about everything they were allowed to take, as to weight . . .
>
> Regarding the weapons: There was a lengthy discussion about weapons. At the end, the Ebens didn't really care. So our team members decided to take some, just in case. Not for a fight, God knows, as our team was vastly outnumbered, but for the safety aspects of it. Remember, the 12 were all military members, so weapons made them FEEL safe. A side note: They only took fifty rounds of ammunition per handgun and one hundred rounds of ammunition per rifle.

EXPLORATION OF SERPO

All the information in this section took up the entire seventh e-mail posting by Anonymous, sent on November 17, 2005.

> *Our team contained two geologists (they were also cross-trained as biologists). The first thing our geologists did was map the entire planet. The first step was to divide the planet in half, thus creating an equator. Then they established a Northern hemisphere and Southern hemisphere. Within each hemisphere, they created four quadrants. Finally, they established the "North and South" Poles. This was the easiest method to study the planet. Most of the Eben communities were placed along the equator. However, there were some communities established north of the equator in each of the four quadrants in the northern hemispheres.*

Plate 1. Photograph of Haunebu IV "Dreadnought"

Plate 2. Artist's rendering of antigravity discs and U-boats at Neuschwabenland, by Jim Nichols

Plate 3. Artist's rendering of the Roswell crash, by David Hardy

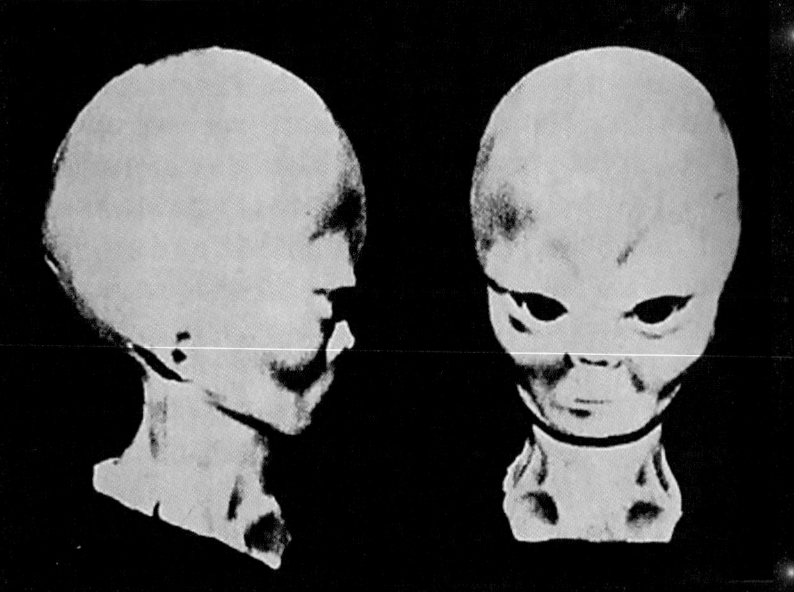

Plate 4. Sculpture of a Reticulan (believed to be identical to an Eben) by Hollywood artist Alan Levigne

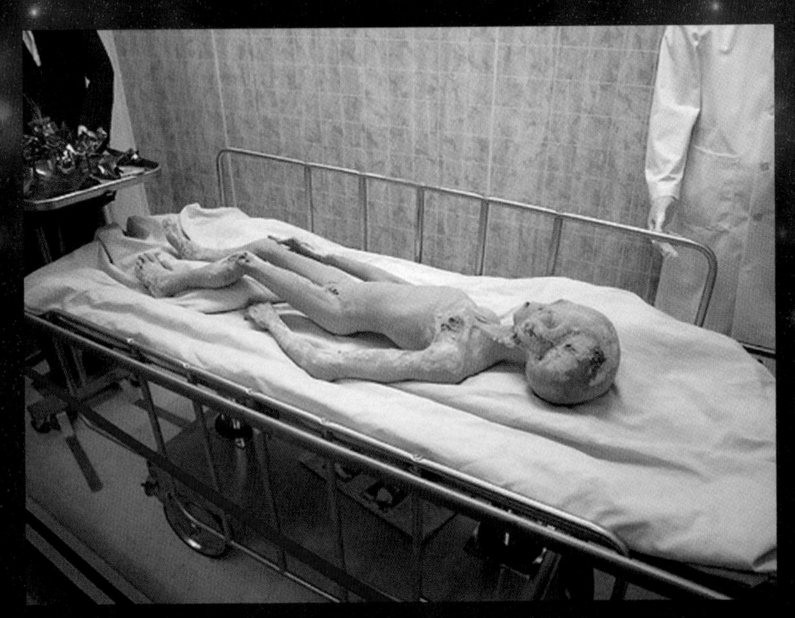

Plate 5. Model of an alien creature found at the Roswell crash site (Roswell Museum, Roswell, New Mexico). This model was used in the movie Roswell, and was donated to the museum by executive producer Paul Davids. It is actually a good likeness of an Eben.

Plate 6. Artist rendering of crashed disc in Kingman, Arizona, by Jim Nichols

Plate 7. Vostok 1 Space Capsule at the RKK Energiya museum

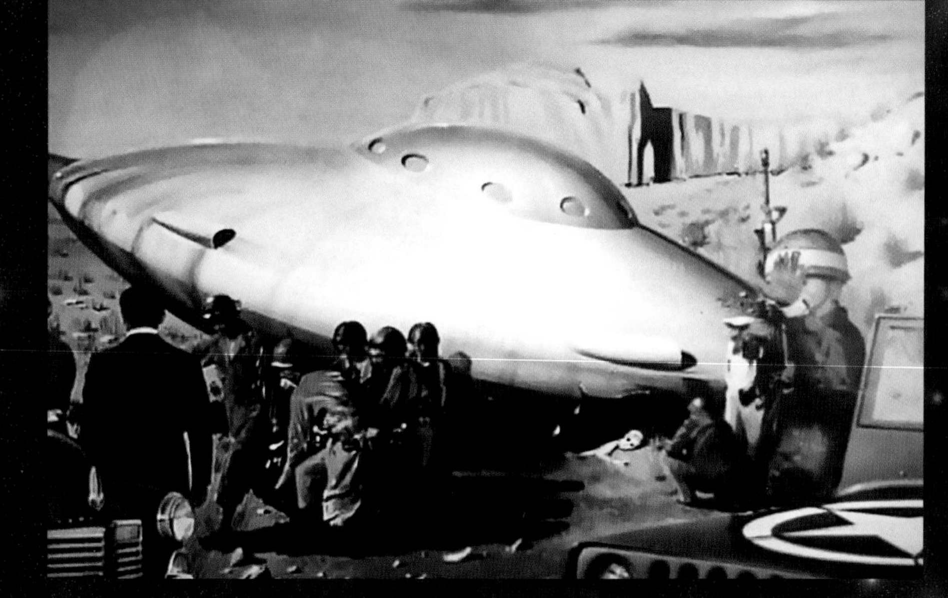

Plate 8. Artist rendering of retrieval of the second Roswell disc near Datil, New Mexico, by Jim Nichols

Plate 9. Alien craft landing scene from Close Encounters of the Third Kind

WORMHOLE IN SPACE
Wormholes would act as shortcuts connecting distant regions of space-time. By going through a wormhole, it might be possible to travel between the two regions faster than a beam of light through normal space-time.

Plate 10. Hypothetical schematic of a traversable wormhole

Plate 12. The alien craft travels through space-time, illustration by Brett Fitzpatrick (www.starbrightillustrations.com)

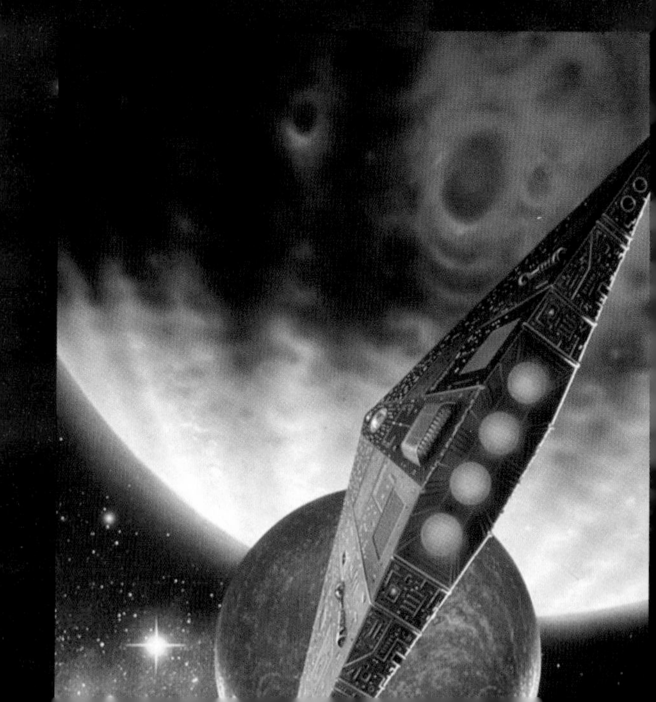

Plate 14. The two suns of Serpo

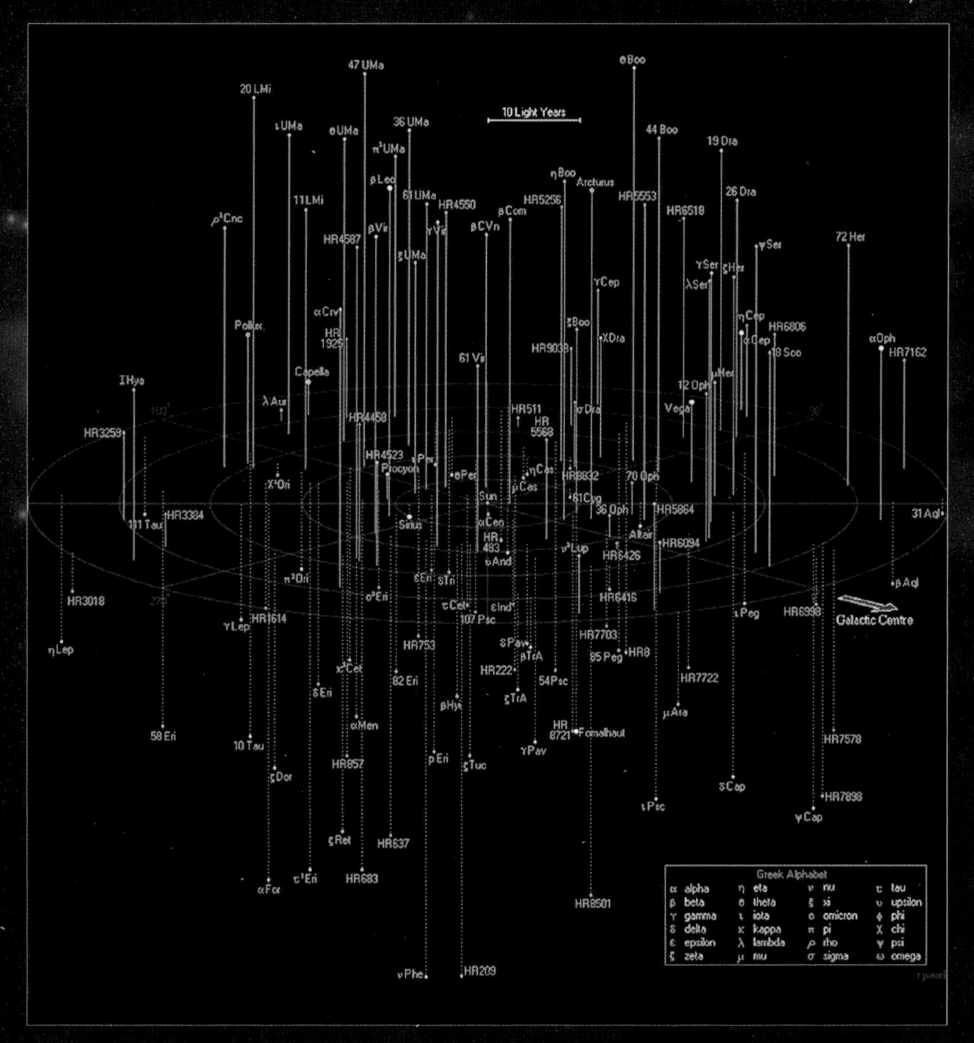

Plate 15. Chart of stars within fifty light-years of our sun. (Zeta Reticuli is tenth from the left in the lower section.)

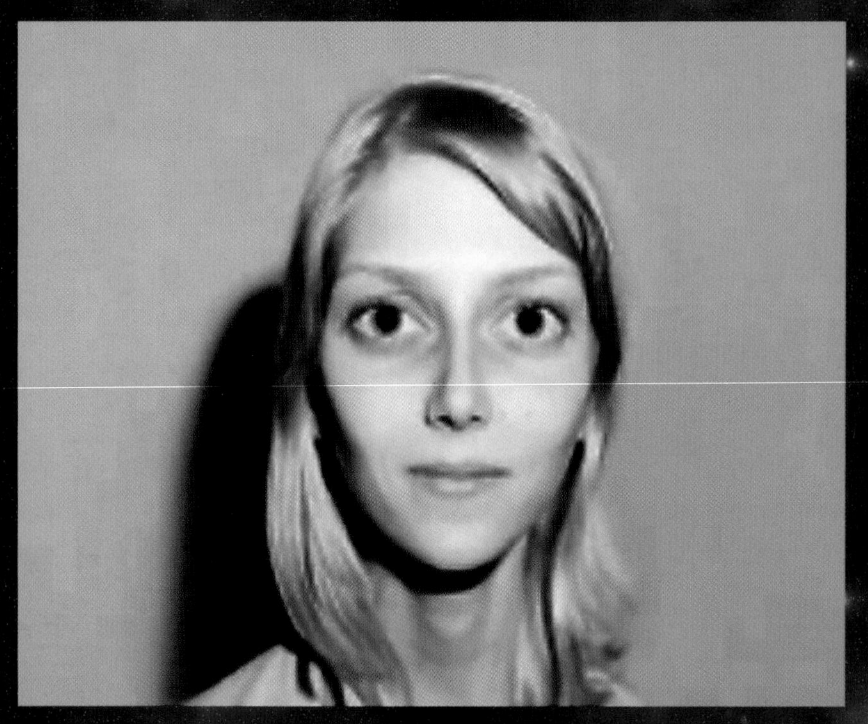

Plate 16. Image of a human-alien hybrid girl, by Andy Social

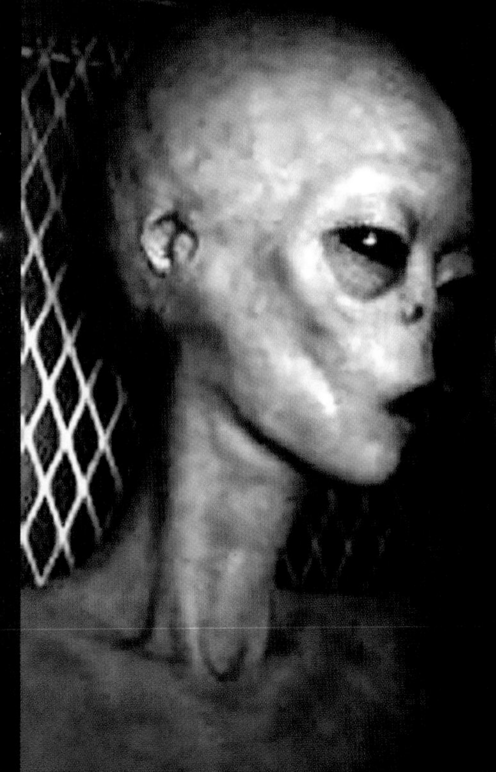

Plate 17. Is this J-Rod? From "Alien Species: Advanced Humans, Greys and Reptilians" by Alton Parish 10/12/12 (BeforeIt'sNews.com)

THE ELECTROMAGNETIC SPECTRUM

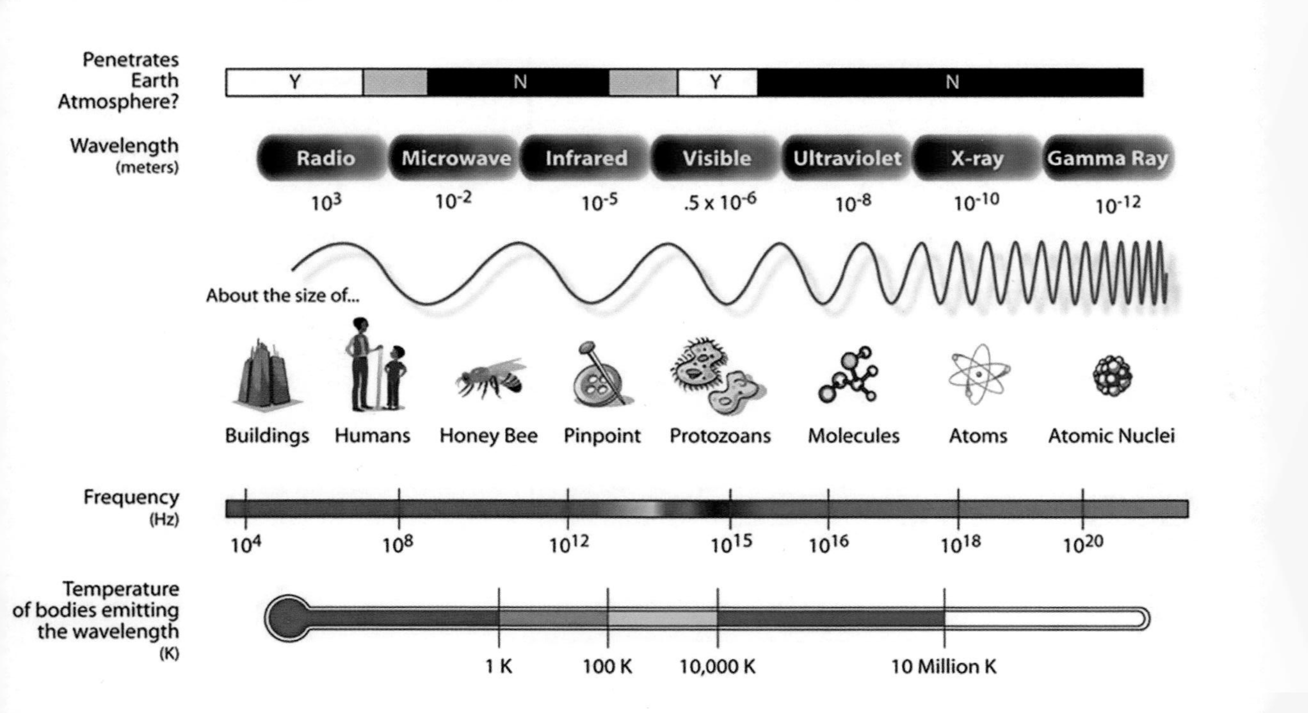

Penetrates Earth Atmosphere?
Y
N
Y
N
Wavelength (meters)
Radio
Microwave
Infrared
Visible
Ultraviolet
X-ray
Gamma Ray
10^{3}
10^{-2}
10^{-5}
$.5 \times 10^{-6}$
10^{-8}
10^{-10}
10^{-12}
About the size of...
Buildings
Humans
Honey Bee
Pinpoint
Protozoans
Molecules
Atoms
Atomic Nuclei
Frequency (Hz)
10^{4}
10^{8}
10^{12}
10^{15}
10^{16}
10^{18}
10^{20}
Temperature of bodies emitting the wavelength (K)
1 K
100 K
10,000 K
10 Million K

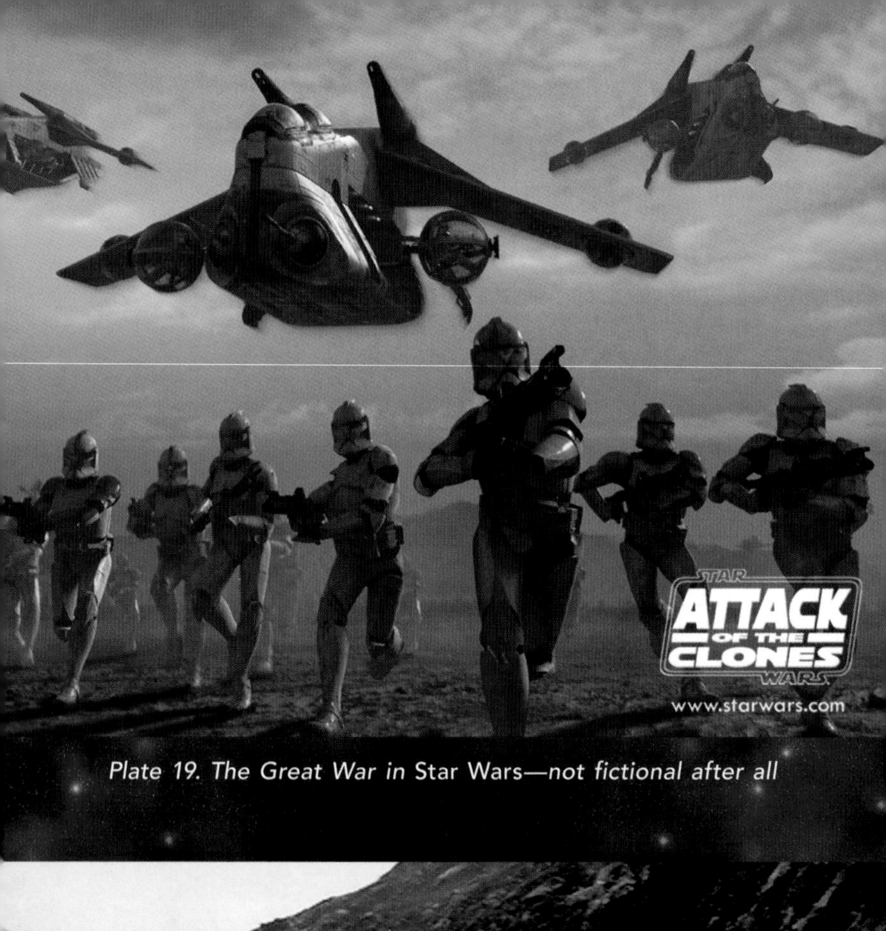

Plate 19. The Great War in Star Wars—not fictional after all

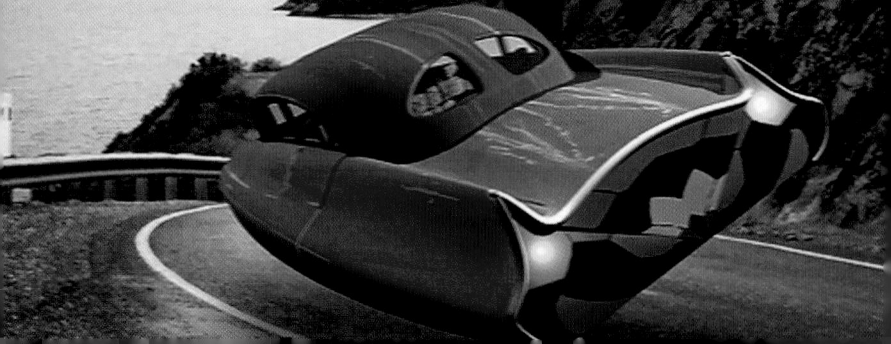

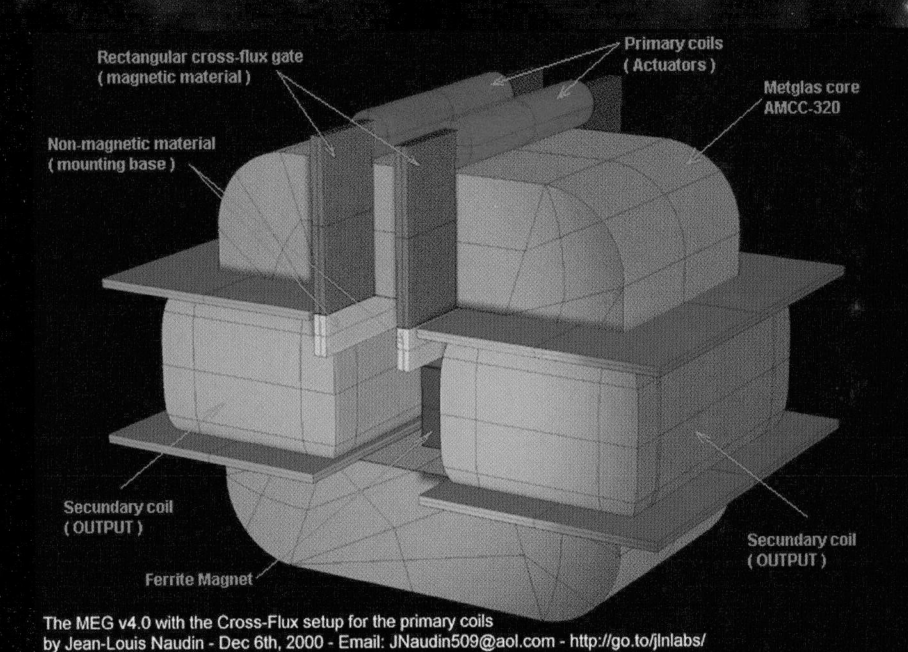

Plate 21. The motionless electromagnetic generator; U.S. patent no. 6,362,718

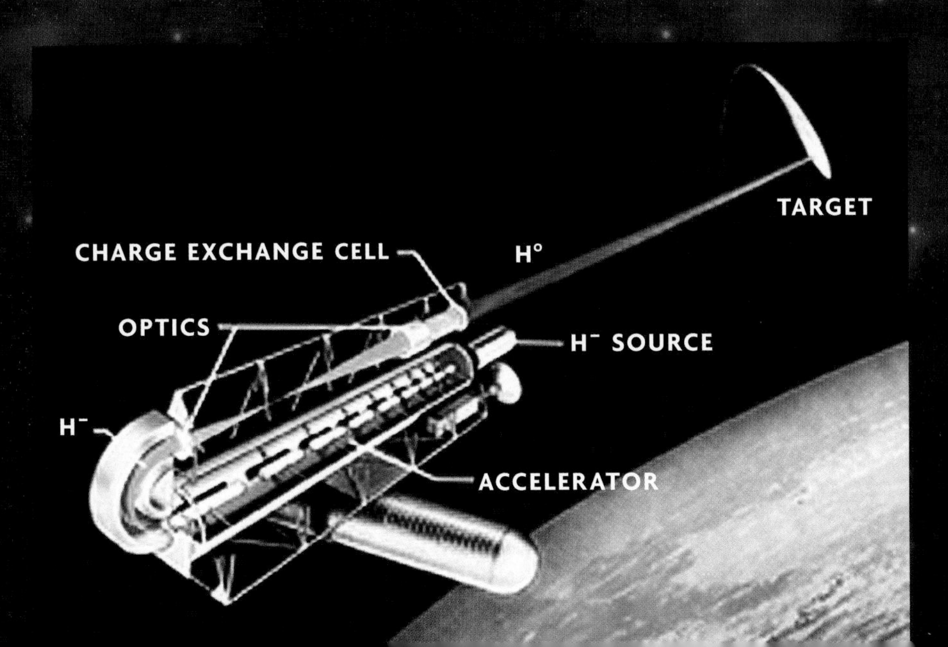

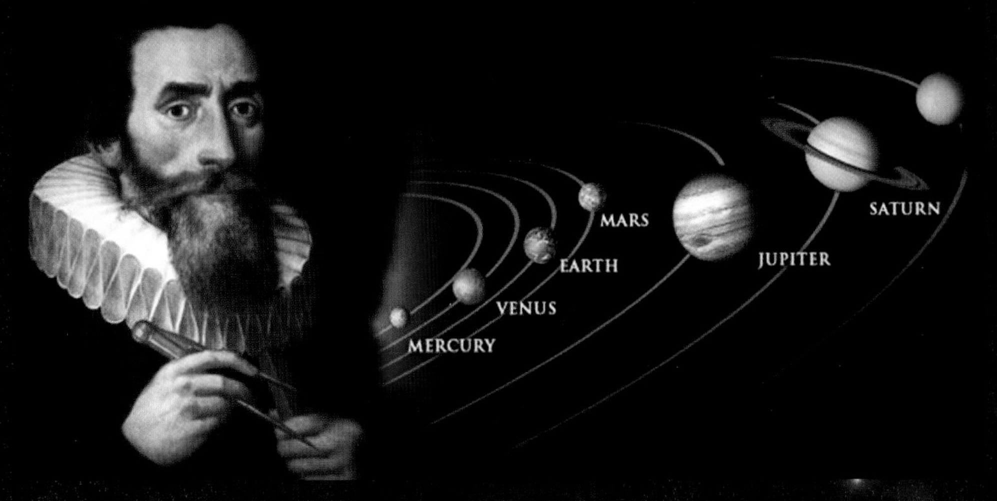

Plate 24. Johannes Kepler (1571–1630)

Plate 25. Montana landscape, similar to Quadrant 1

Plate 26. Monsignor Corrado Balducci (1923–2008), chief proponent of Vatican belief in alien visitation

Plate 27. The Chronovisor, a device for seeing past events, developed by two Italian priests, image copyright Lee Krystek

There were no communities located in either poles. The southern pole was desert. It was barren land with virtually no precipitation where absolutely nothing grew in this area. . . . There were volcanic rock formations and part of the extreme south contained a rock desert. Temperatures in the south pole were measured at between 90°F and 135°F. Going further north from the south pole in Quadrant 1, the team found extrusive rocks. This indicated some volcanic activity in the area. Our team found numerous volcanoes in this area. The team found several fissure eruptions in this region, with standing water. The water was tested and contained high levels of sulfur, zinc, copper and other unknown chemicals.

Moving from east to Quadrant 2, the team found basically the same volcanic fields of rocks. However, in one particular location near the north end of Quadrant 2, the Team found an Alkali Flat. On Earth, these were formed by streams flowing into a desert or arid location. Our team found hard mud covered by alkaline salts. Some vegetation was growing in this area. Moving to Quadrant 3, the team found a form of Badlands: an arid region that is lined with deep gullies with sparse vegetation. The gullies or valleys were extremely deep, some going down 3,000 feet. The team found the first Serpo animal in this region. It looked like an armadillo. This creature was extremely hostile and tried to attack the Team several times. The Eben guide used some type of sound device (sonic-directed sound beam) to scare away the creature.

Moving to the equatorial region, our team found desert-style landscapes which contained patches of vegetation. The team found numerous pockets of water fed to the ground by Artesian Wells. This water was the freshest, containing only the unknown chemicals. It tasted good and the Ebens drank and used it. Our team still boiled it because during culture tests, unknown types of bacteria were detected. Moving to the Northern Hemisphere, the team found a major change in climate and landscape. One Team member, who had coined it Quadrant 1, in the northern hemisphere, named it "Little Montana." The Team found trees, similar to the Evergreen style of Earth trees. These trees were milked by the

Montana landscape, similar to Quadrant 1 (also see plate 25)

Ebens. A white fluid was extracted and drank [sic]. Numerous other types of vegetation was [sic] found in this region. Standing water, possibly fed by either Artesian Wells or fissure eruptions was found. In one area, marshlands were seen. Large plants were observed growing in the marshy area. The Ebens used these plants for food. The bulb of the plant was very large. The bulb tasted something like a mellon [sic].

THE MOVE TO QUADRANT 1

Our team eventually moved to an area in Quadrant 1 in the northern hemisphere. This area contained moderate temperatures [50°–80°] and ample amounts of shade. The Ebens built a little community for the team. Most of the remaining exploration of the planet was done from this point. The Team only explored the southern hemisphere once, obtaining geological information. Because of the intense heat, the Team decided not to venture back. The Team continued to explore the northern hemisphere, where, as *the Team traveled towards the north pole, the temperature cooled considerably. The Team found mountains, raising to an elevation of 15,000 feet and valleys that sank below the basic mark the team established for "sea level." Lush green fields were found containing a form of grass, but contained bulbs. The team coined these fields "Clover Fields" even though the bulbs were not clover. The radiation levels were lower in the northern hemisphere than at the equator and southern hemisphere. The north pole contained cold weather and the team saw the first sign of snow. Blankets of snow littered the landscape around the north pole. The snow measured about 20 feet at its deepest. The temperature was a constant 33°. Our Team never found the temperature to vary in this region.* The Ebens could not stand to be in this region for long. They suffered extreme hypothermia. The Team's guide wore a suit, similar to a space suit with built in heaters.*

Our team found evidence of past earthquakes. Fault lines were found along the northern tip of the southern hemisphere. Exfoliation was observed along with extrusive rocks, which indicated magma flows in the past. Our Team brought back hundreds of samples of Serpo soil, vegetation, water and other items for testing on Earth. During our Team's exploration, they discovered numerous types of animals. The strangest was the "Beast" which looked like a large Ox. The animal was timid and never seen to be hostile.

Another animal looked like a Mountain Lion, but had long fur around the neck. This animal was curious, but was not considered hostile by the Ebens.

During the exploration of Quadrant 4 of the southern hemisphere, the team found a very long and large creature that appeared to be a snake. This creature was "deadly" as explained by the Ebens. The head of the creature was large and contained almost humanlike eyes. This was the only time our Team used their weapons and killed the creature. The Ebens didn't appear to be

*It wouldn't seem likely that twenty feet of snow would remain intact in 33° temperatures. However, this might be explained by the indirect sunlight at this latitude.

Military-issue, Vietnam-era Colt .45

upset that the Team killed the creature, but was upset they used a weapon. The team brought four .45 caliber Colt (standard military issued) handguns and four M2 carbine rifles. After killing the creature, the team disected it. The internal organs were strange and nothing similar to a Earth-style snake. The creature measured 15 feet long and 1.5 feet in diameter. The team was curious about the eyes. Examination of the eyes revealed cones, similar to [those found in] human eyes. The eye contained an iris and the back contained a large nerve, similar to the optic nerve feeding into the creature's brain. The brain was large, much larger than any Earth-based snake. The Team wanted to eat the meat from the creature, but the Eben guides politely told them "No."

The bodies of water on Serpo did not contain fish, as we know [them]. Some bodies, near the equator, did contain strange-looking creatures, similar to eels (small and about 8–10" long) and was probably a cousin of the land-based "snake." There was something like a jungle, near the marshlands, but not the jungles that we're familiar with.

COAST TO COAST

The following posting by Anonymous in Release 10a on December 8, 2005, was in reaction to having listened to a *Coast to Coast* radio interview of Richard (Rick) C. Doty, a retired Special Agent of the Air Force Office of Special Investigation (AFOSI), and Bill Ryan, the Serpo website creator, about Project Crystal Knight the previous evening. Information in brackets was added by Victor Martinez.

> Yes, I did listen to the entire show. Never heard this guy George Noory before, but he seemed to be a very open-minded narrator and for that, me and my DIA colleagues were very pleased. The show caused a lot of buzzing in the halls of the DIA! Bill [Ryan] and Rick [Doty] both did an excellent job. I was hoping you [Victor Martinez] might just show up and add your own comments since you know more about me than anyone else as well as the program, but I guess that wasn't possible. I called Mr. _____ this morning ***[EX U.S.G. OFFICIAL WHO IS MANAGING THE CONTROLLED RELEASE OF "Project SERPO" FOR PUBLIC CONSUMPTION].*** I found some discrepancies about the animals mentioned on the show. The Armadillolike creature was not aggressive; it just scared the Team Members. The Eben guide directed some type of sound (very highpitch) at the Armadillo creature and that scared it away. These creatures were seen at several locations around the planet. Some were larger than others, but they were not aggressive. Only the snakelike creature was aggressive, which forced the team to kill one. The snake-like creatures were located in just one location and the Team never saw another one. As for birds, there were two types of flying creatures. One resembled a hawk and the other looked like a large flying squirrel. Neither were aggressive and the Team could never catch one for examination. As for insects, they had small bugs, similar to cockroaches, but smaller. They were harmless, but did get into the Team's equipment. They had a hardened shell, with a soft interior body. The team never observed any flying insects, such as flies, wasps, etc. Several other small bugs were found and identified.

RADIATION

The Team Commander knew that they were absorbing high levels of radiation, and spoke about it in a diary entry previously quoted. A Comment received on November 15, 2005, discussed the radiation problem.

> One thing that concerns me is the radiation the team was receiving. I would think the team had some sort of radiation detection equipment. Back then, they had Radacs. We also had dosimeters. I'm sure each team member wore one and must have read it periodically. Knowing this was a military mission, someone must have been designated a health monitor, probably the physicians. If the team realized they were receiving large doses of radiation, why didn't they bring this to the Ebens attention. If the Ebens were so benevolent, then they might have provided some form of protection to the team members or provided them with an antiradiation pill or something. Or, once the team received the maximum dose, why didn't the Ebens bring them home? One of my former colleagues thinks the radiation could be something different, in another wavelength and thus our team members didn't register that particular radiation dose.

Modern-day dosimeter

The fact that the radiation level was lower in the Northern Hemisphere was another reason they decided to remain in Quadrant 1. It makes sense that radiation levels would be lower in a location that received less direct sunlight. The team stayed in Quadrant 1 for the remaining seven years of their sojourn on Serpo. However, that didn't solve the radiation problem. Anonymous says:

> Each team member received a large dose or [*sic*] radiation during their stay on Serpo. Most of the Team Members died later of radiation-related illnesses.

16
THE RETURN

The team remained on Serpo for thirteen years. They paid a three-year penalty for their inability to keep track of Earth time. After all their time devices failed, they tried desperately to translate one revolution of Serpo to one revolution of Earth, but were not able to keep up a counting discipline.* They knew that a Serpo day was 43 Earth hours, so to compute Earth months from Serpo months they would need to multiply 43 times 30.2 and divide by 24. That means that 54.11 Earth days passed for each Serpo month. So, all they had to do was to keep track of Serpo months of 30.2 Serpo days each, and then multiply by 54.11 to calculate the Earth time passage in days. But, somehow, their record keeping broke down and they couldn't reset their results properly. It's easy to understand how this could have happened, despite their best-intentioned plans to maintain military discipline. Marking Serpo days by the movements of the two suns may have been more difficult than it seems. And then, of course, how does one accurately measure 20 percent of a day?

SEVEN RETURNED

The team returned to Earth on August 18, 1978. Only seven team members returned.† Three had died, one in the alien spaceship en route

---

*Serpo revolves around the Zeta 2 star. See appendix 4 for details.

†There is confusion on this point. Anonymous states in several places that eight team members returned. However, that conclusion cannot be justified, since we know from the Team Commander's diary that three people died.

to Serpo and two on the planet. Two decided to remain on Serpo and to live out their lives there. Anonymous responded to a question about why two team members decided to stay. He says:

> Why some team members remained! The debriefing reported that the Team Members who remained did so voluntarily. They fell in love with the culture of the Ebens and the planet. They were not ordered to return. Communication with the remaining crew members [the two who stayed on Serpo] lasted until about 1988. No other communication was received from those team members. The two who died on the Planet Serpo [899 and 754] were placed in coffins and buried. Their bodies were returned to Earth.*

Regarding the debriefing, Anonymous has this to say:

> The returnees were isolated from 1978 until 1984 at various military installations. The Air Force Office of Special Investigation (AFOSI) was responsible for their security and safety. AFOSI also conducted debriefing sessions with the returnees.

A comment sent in to the website by someone who calls himself Anonymous II adds the following:

> We did have a special unit that handled their debriefing but USAF positive intelligence was also involved. I was never involved in that program, but I knew other agents who were.

There is common agreement that the returning team was debriefed for a solid year, and the information they gave is contained in a three-thousand-page document. Anonymous claims to have access to that

*This, of course, doesn't make sense. It seems unlikely that they would have buried and then disinterred the two bodies later, when, given their bioscience sophistication, they probably had a way to preserve them, as we did with the Eben bodies. They knew that the team was departing in 1975, so they would just have had to plan to preserve their remains for about seven years, although it turned out to be ten years.

document, and to have garnered all the information that was submitted to Victor Martinez, the e-mail moderator, from that book. Rick Doty, commented on that in an article he wrote for *UFO Magazine* in 2006.* Doty says:

> We must remember that Mr. Anonymous will hardly have the 3,000-page report in his living room just sitting there like a Sears catalog. The report will be guarded under the tightest conceivable security and the conditions of access are unknown by us. We can hypothesize that Mr. Anonymous may not even have access to the documents at all and may be relying on memory, someone else's memory, or someone else supplying him with the information maybe by phone or by tape under conditions over which he himself has no control.

This might explain some of the inconsistencies. However, Anonymous is highly qualified as a historian on ET contact affairs. See the Introduction for a detailed discussion about his involvement with the *Red Book,* a complete history of government-extraterrestrial contact. According to Anonymous, the last survivor of the entire team, including the alternates who remained on Earth, died in 2002 in Florida.

THE CREW LOGS

In August 2006, Bill Ryan, the webmaster, received the following message from an anonymous source in U.S. intelligence, and he posted it on the website on August 30, 2006:

> Every single bit of information from the postings regarding the crew logs is absolutely correct. I've verified that the logs are real and were

*Doty is generally considered an insider on all secret UFO information. He was a collaborator with Robert M. Collins, a former Air Force intelligence officer, in the writing of the book *Exempt from Disclosure: The Black World of UFOs,* now in its third edition (Peregrine Communications, 2010). Collins was formerly the chief analyst in theoretical physics at Wright-Patterson AFB in Ohio, in the Foreign Technology Division. This book gives reliable inside information about several top secret Air Force programs, including some of the projects carried out at Los Alamos National Laboratories.

transcribed from official tape recordings made by the crew. There are 5,419 cassette tapes that contain voice recordings. I've heard one of them, but saw all of them in a secure environment.

I know these logs are genuine. I was given the opportunity to see them, and listen to one in its entirety, which was made by the Team Commander himself.* How could someone fake 5,419 × 90 minute tapes? Do the math: it would have taken someone 338 days to fabricate the tapes. That type of government cassette tape is no longer sold today, but they were military issue then. There were no cassette tapes available to the public until 1968. The crew took 60 boxes each containing 100 × 90 minute cassette tapes, which is 6,000 tapes.

These tapes were recorded during the mission to Serpo, over the course of their years away from Earth. Each crew member had a recording device and recorded their observations. Once they returned, the tapes were transcribed over a period of seven years.

Look at the critics trying to destroy this fantastic true story. If I were the keeper of this information, I wouldn't release it to anyone . . . except the general public, through the news media. There are some true disbelievers in the UFO community who wouldn't believe the story if God himself told it.

THE PHOTOGRAPHS

The story of the release of the photographs taken on Serpo is tortuous. Victor Martinez did everything in his power to obtain and publish the best one hundred of the over three thousand photos brought back to Earth by the team. But despite the tireless and heroic efforts of both himself, Bill Ryan, and Anonymous, they were blocked at every turn. Ultimately, they had to concede defeat. Anonymous did eventually send Martinez a computer disk containing six photo images, but as will be seen, five of those were compromised, and in the end, only one

---

*We have seen previously that some of the Team Commander's diary entries were handwritten. Evidently, he used both methods for recording his notes. The handwritten entries were made on the first and second days of the mission. More than likely, he switched over to the cassette tapes shortly after that.

questionable photo survived. The forces preventing the publication of the photos were just too powerful.

Following is a shortened time line of events surrounding the release of the six photos sent by Anonymous, consisting of extracts of direct quotes from Release 22, written by Victor Martinez. It is reproduced here to give the reader a taste of the complexities surrounding obtaining and reproducing the photos, and to counter the inevitable objections to the Serpo story, based on failure to release the photos.

> Late Saturday afternoon [December 9, 2006], I go to my PO Box and there's a CD waiting for me in a nondescript package. I knew instantly what it was and I went directly to Kinko's to view it. Since I don't even know how to insert a disk into a computer, I immediately "hire" one of the employees to open it for me and execute any and all functions I want/need; I offer him $20/hour above his hourly wage. Each image/file was 13.946 megabytes in size and could never have been received by me at WebTV which has an incoming capacity of 10 megabytes total. I had him make two (2) sets of color prints for me and then quickly left; I informed him and other curious employees that what he/they had viewed were pictures from a future sci-fi movie produced by SKG [Spielberg Katzenberg Geffen] in Hollywood.
>
> **Monday, Dec. 11, 2006:** I overnight the disk to a retired signals intelligence data/photo/ digital analyst for the National Security Agency [retired in Linthicum, MD] . . .
>
> This SERPO PHOTO RELEASE had been in the works since September 2006. I only have "The Suns Set" photo on hand to share w/ Bill [Ryan] and Kerry [Cassidy] at the time.* Bill provides me something called the FTP for the SERPO.org website, so I inform him that I'll have my NSA contact friend download the images directly into his website.
>
> **Thursday, Dec. 14:** My NSA contact friend phones me around 6 p.m.

*This photograph appears to be the only authentic Serpo photo that has survived all the complex negotiations. The fact that the lower left-hand corner is blacked out indicates that someone did not want the details of the photo shown, and testifies to its authenticity. It is shown here and in color plate 14.

"The Suns Set" photo (also see plate 14)

to [regrettably] inform me of a huge slew of problems w/what ANONYMOUS' "Gatekeeper" has sent me. I'm floored, EXTREMELY DISAPPOINTED and dejected by what he has to say; four hours later we're finally off the phone.

Friday, Dec. 15: I mail a dup copy of the CD w/ the original UNcompressed images to Bill at 10:24 a.m.; I have NO copy(ies) for myself.

Friday, Dec. 15: At ~8 p.m., all of the images [compressed] have been successfully downloaded into Bill's SERPO.org website by my NSA friend.

Sunday, Dec. 17: Right after Sunday brunch, Bill shocks me w/ a cell-to-cell call placed at 1:30 p.m., informing me that his SERPO.org website has been successfully HACKED into and that two (2) of the photos have been posted on a DEbunking Web site! They're "outed" as FAKES—something Bill, my NSA contact, and myself already had concluded—and we were NOT going to present them as being authentic/genuine . . . ever. In a weird sort of way, this CRIMINAL ACT actually assisted us! It's important to note that the images obtained by the COMPUTER CRIMINAL (probably from the UK) were NOT the original ones that were received by me from "The Gatekeeper"; they were compressed JPG files and we knew instantly that the source for these photos was Bill's website because of another built-in security feature: The images on the ORIGINAL CD mailed to me have completely different file names; the ones downloaded from Bill's website were REnumbered by my NSA contact to coincide with the stream sequence. THAT's how we KNOW they were ILlegally [*sic*] obtained via Bill's website.

We finalize plans on the time of the stream and Bill advises me that after having viewed the images for himself (remember, he hadn't viewed them yet at lunch on Thursday), he chooses to reverse his position and WITHHOLD from public view/mask from public view ALL five (5) of the images pending a further clarification from ANONYMOUS. NOTE: Only "The Suns Set" is being made available at this time as "94 of 100." I tell him that I fully support his decision . . . whatever he chooses to do at this point; I trust, support, and reaffirm his sound judgment.

Subject: REPORT BY RETIRED SIGNALS INTELLIGENCE ANALYST, NATIONAL SECURITY AGENCY, Ft. Meade, MD / PHOTO ANALYST REPORT ON SIX SERPO PHOTOS, 17 Dec 2006 RE: Conversion of Six Images to .JPG Format

EXECUTIVE SUMMARY: Six images in digital format were provided for reprocessing. The conclusion by the tasked analyst is these six images require more investigation, analysis, and verification, and lack credibility in many technical respects. They are far below NASA/JPL [Jet Propulsion Labs] 1960s standards in terms of quality. This mixture of six images, taken as a group, is inadequately explained. There is low confidence that these six images are related to a real Serpo disclosure.

Details of image processing: The original six images were supplied in .BMP format under the file names: "serpoimages0001.bmp" through "serpoimages0006.bmp." The files as received were not encrypted. The original image file names were changed during creation of the .JPG files, to fit an expanded numbering plan for internal reference purposes, in case several sets of images with the same names were received over time. Markings indicated that the original images from this set were in

"Volume 12, Section 24" [Bolling Air Force Base, "Project SERPO" case files].

All images (including any excess white area) were 1700 x 2800 pixels and approx. 14.3 megabytes each in size. For display and retrieval purposes on the Internet, the images were resampled to a width of 800 pixels with a proportional height, ranging from 427 to 935 pixels. A 2% reduction of red was applied to color-compensate for conversion to nonprogressive .JPG format at 15% compression.

Except for cropping to eliminate excessive white space, the pictures were deliberately not improved using software filters and standard techniques. The original source image for #97 was upside down. It has been rotated 180 degrees. The original six .BMP images have not been subjected to detailed steganographic analysis. There is evidence of moiré patterns in these images, particularly #94a and #94b. Typically, this is an artifact of scanning a color halftone picture. These patterns or interference lines are much more pronounced in the original .BMP files than in the smaller .JPG versions.

Such moiré patterns are inconsistent with direct scanning of 35mm photographic film or continuous-tone prints to digital format. In contrast, the moiré patterns (and telltale irregular black borders around images #98 and #99) are consistent with pictures roughly cut out and scanned from books, magazines, and other printed sources.

—RETIRED NSA ANALYST

SUBSEQUENT EBEN VISITS

In Release 32 by Anonymous, he gives the past and future dates of Eben visits to Earth. This confirms the 1978 return of the team, thirteen years and one month from the date of departure. The first six return visits were to the Nevada Test Site (NTS).

The EBEN visits from ZETA RETICULI I and II occurred/will occur on the following days and dates:

Friday, August 18, 1978
Thursday, April 28, 1983
Sunday, April 7, 1991

Tuesday, October 22, 1996

Sunday, November 28, 1999

Wednesday, November 14, 2001

Thursday, November 12, 2009 at the U.S. unincorporated territory of JOHNSTON ATOLL on AKAU Island

THE LANDING AT JOHNSTON ATOLL

Last month, on Thursday, November 12, 2009, a visit WAS made by the Ebens to a remote location on Earth, but it was NOT to the NTS, but instead to the little-known North Pacific Ocean desert island of JOHNSTON ATOLL/ISLAND. The Ebens paid us a kindly visit for 12 hours between 0600 - 1800 on this U.S. territory [U.S. military time, local]. The meeting specifically occurred on AKAU ISLAND, which is the North Island of the sprawling JOHNSTON ATOLL system. The Ebens landed on a flat portion of the northwest sector of this island. There were 18 (eighteen) officials from around the world who met the EBENS. They included the following representatives from our planet, Sol III:

Johnston Atoll in the mid-Pacific

1 = The VATICAN

2 = The UNITED NATIONS

9 = The UNITED STATES of AMERICA

Broken down into

1 = WHITE HOUSE REPRESENTATIVE from the OBAMA ADMINISTRATION

2 = U.S. INTELLIGENCE OFFICERS

1 = LINGUIST

5 = U.S. MILITARY REPRESENTATIVES

Other Countries / Misc.

1 = The PEOPLE's REPUBLIC of CHINA

1 = The RUSSIAN FEDERATION

4 = INVITED SPECIAL GUESTS

= 18 TOTAL INVITED GUESTS

Additionally, a very special set of gifts were exchanged. The Ebens provided us with six (6) gifts that would assist us in future technological developments. In return, the Vatican gave the Ebens two (2) 12th Century

Monsignor Corrado Balducci (1923–2008), chief proponent of Vatican belief in alien visitation (also see plate 26)

> religious-themed paintings. Furthermore, a future meeting date was set for November 2012 in addition to a previously scheduled official visit on Thursday, November 11, 2010 at the NTS.

We know now that the Ebens did indeed return to the NTS on November 11, 2010, and in November 2012 (date unknown) as planned. Victor Martinez gives the following information about Johnston Atoll:

> JOHNSTON ATOLL is a 50-square-mile atoll and is one of the MOST ISOLATED ATOLLS IN THE WORLD, located in the North Central Pacific Ocean, some 750 miles southwest of Honolulu; the nearest lands to JOHNSTON ATOLL are the tiny islets of the French Frigate Shoals, around 800 km to the northeast, in the Northwestern Hawaiian Islands. It is about one-third of the way from Hawaii to the Marshall Islands. Located at 16° 45' North LATITUDE and 169° 30' West LONGITUDE, there are four (4) islands located on the coral reef platform, two natural islands—JOHNSTON ISLAND and SAND ISLAND—which have been greatly expanded by coral dredging, as well as North Island (AKAU) and East Island (HIKINA), both of which are artificial islands formed solely by coral dredging. JOHNSTON ISLAND is an unincorporated territory of the United States, administered by the U.S. Fish and Wildlife Service of the Department of the Interior as part of the Pacific Remote Islands Marine National Monument managed by the National Wildlife Refuge System. The defense of JOHNSTON ATOLL is managed by the U.S. military; NONE of the islands are open to the public. It is a former ATOMIC TEST FACILITY and site for five (5) high-altitude nuclear tests during the 1960s called "STARFISH PRIME" grouped together as "Operation FISHBOWL" under the larger former classified project of "Operation DOMINIC."

THE FINAL REPORT

The seven surviving Serpo team members returned in August of 1978. We know that they were debriefed for a solid year following their arrival back on Earth, resulting in a 3,000-page book containing all their reports. Then, in 1980, a Final Report was written. As previously mentioned, Carl Sagan was one of the signatories to that report. The Final Report is being kept in a vault at DIA headquarters at Bolling Air Force Base in Washington, D.C., along with the debriefing book, all other Serpo-related documents, audio recordings, and photographs. For purposes of assisting future historians who may be able to use the Freedom of Information Act to gain access to all this material, it is important to identify the federal document number of the Report. There is some dispute about that number. Gene Loscowski (see Introduction and appendix 2) says it is 80HQD893-20. Paul McGovern (see Introduction) agreed and added that it was classified T/S (Top Secret), Codeword. This number was confirmed by Victor Martinez, who claimed it to be the same as the number received from Anonymous. Moreover, this is the identical number displayed in the title frame of the *Film From the Box* video discussed in appendix 13. So there is every reason to believe that this is the actual Final Report number. However, in a Comment submitted to the website, the author, who said "Most of the released information is truthful," said also, "The final report is contained in a document entitled "QW." The title was classified. The document number is: #80-0398154." While the overwhelming evidence supports the previous number, this is reported simply in the interest of giving future researchers everything that was submitted.

And so, that closes the final chapter of this remarkable story. We know that the last survivor of the Serpo adventure died in Florida in 2002. Their lives were shortened by the intense radiation they absorbed on Serpo. They died as they lived, "sheep dipped" and unknown to the world. We sincerely hope that they weren't made to suffer the final indignity of not having their real names engraved on their tombstones. We would hate to think that in death, as in life, they remained identified only by three-digit numbers. We will remember some of them as "Skipper," "Sky King," "Flash Gordon," "Doc 1," and "Doc 2." Regardless, they are now clearly engraved in history as giant figures in the inexorable march across the galaxy by the human race.

PART THREE

EPILOGUE

What we learned from Eben science and technology essentially revolutionized our current understandings of physics and biotechnology. The fact that these understandings have not yet "trickled down" to earthly industries is due more to matters of international relations, politics, and economics than to the readiness of our universities, corporations, and medical institutions to translate this amazing knowledge into practical usage. Our scientists were already beginning to comprehend this new paradigm of the universe in such areas as quantum physics and advanced electromagnetics, and our biologists were close to breakthroughs in DNA biotechnology. So, we could be utilizing Eben science today and could be living in virtually Utopian conditions. It is inspiring to know that these amazing ground-breaking innovations are waiting for us just around the corner, once we resolve our tangled earthly affairs. We are already flying antigravity craft and are using stem-cell technology to cure a host of previously intractable medical conditions. This is just the beginning. In this part of the book, I have summarized the main categories of Eben technology.

While these advanced technological breakthroughs are being kept from us, there is one source of futuristic information that they cannot control. We can, at least, learn about what an incredible lifestyle awaits us from science-fiction film. These movies energize our imaginations and sustain us while we wait. They have allowed us to live in the new world in unreality until the reality arrives, especially since we now understand that they may be part of the Public Acclimation Program, and are actually true depictions of what already exists. I believe that *Stargate,* released in 1994, was a good example of this.

Steven Spielberg's classic film, *Close Encounters of the Third Kind,* released in 1977, gave movie audiences a thrilling taste of the future as it showed twelve American military personnel boarding an alien craft and departing our planet for a distant star system. While we all thought it was fiction, there was something about that scene that made us believe it could and would happen. And, with the revelations about Project Serpo that emerged twenty-eight years later, we know now that it was real. With that in mind, I thought it instructive to go back and analyze that film and compare it to the reality to comprehend how Hollywood sci-fi movie-makers may be showing us the future in many other films that also now may seem fantastic. The recent very successful film *Avatar* comes to mind. More and more we may now consider these films to be sci-fact, as was *Close Encounters.* How filmmakers have been able to divine the future in such detail must remain mysterious until we all gain the same ability and time, as we have known it, no longer exists.

17

EBEN TECHNOLOGY

Some of the more notable examples of the advanced Eben technology have already been discussed in other chapters. However, Anonymous mentions only briefly some other remarkable technologies, and does not give details, usually devoting only a line or two to these cases. It would be helpful here to list these other technologies, and to speculate as to how they might work. Our government has, of course, wanted the Ebens to share this information with us, and more than likely, since the aliens are very open and cooperative, they may have already helped us to develop these technologies. It would not be at all surprising to find out that we do indeed already have these capabilities. Certainly, the work on these projects would be hidden in "black projects"; that is, in underground facilities and in classified laboratories of select universities and corporations. These other Eben technologies mentioned by Anonymous earlier are listed here with appended known information and/or speculation.

1. Free energy. Anonymous says, "They were able to tap into a vacuum and bring back an enormous amount of energy from that vacuum." We are not new to this research and development. In the early 1920s, Dr. T. Henry Moray, who was a boyhood admirer of Nikola Tesla, built a free-energy device, weighing about sixty pounds, with no apparent power input, that produced fifty thousand watts of electricity for several hours. Despite all the publicity surrounding this device, the U.S. Patent Office never granted him a patent. Moray met with constant opposition. He was frequently threatened; his laboratory was broken into and robbed, and he was actually shot at in the street and in his own office. More recently, physicist Thomas E. Bearden has been working on it, and has developed the MEG, or motionless electromagnetic generator (see plate 21). It has no moving parts. It was granted a U.S. patent on March 26, 2002. According to Bearden's website, he has recently achieved a 100:1 ratio of energy out to energy in. Those interested should read Bearden's 977-page magnum opus, *Energy from the Vacuum* (Cheniere Press, 2004).

2. Particle-beam weapons. Anonymous tells us that the Ebens used particle-beam weapons in their hundred-year war, as did their enemy. The Ebens prevailed and destroyed the enemy civilization. These weapons are undoubtedly akin to the famous teleforce ray of Nikola Tesla, which the popular press called the "death ray." They are in the class of directed-energy weapons, which project a stream of ions or electrons at tremendous velocity approaching the speed of light, with each individual ion packing up to a billion volts of power (see plate 22). In his treatise, Tesla said that his particle gun would "send concentrated beams of particles through the free air, of such tremendous energy that they will bring down a fleet of 10,000 enemy airplanes at a distance of 200 miles from a defending nation's border and will cause armies to drop dead in their tracks." The concentrated beam would disrupt the molecular structure of its target. Sandia National Laboratories is already working on such a device in their Ion Beam Laboratory at Kirtland Air Force Base in New Mexico.

3. Antigravity land vehicles. The Team Commander mentions in a diary entry that the Ebens used vehicles that moved above the ground, and therefore had no wheels. We have already learned that the Ebens had the ability to render heavy equipment weightless when they "floated" forty-five tons of team supplies and equipment into their shuttle craft in a single move. We have already developed antigravity aircraft, and it would not seem to be difficult to adapt this technology to land transportation once it is released to the American automobile industry. The principles are the same. Moviegoers are used to seeing these vehicles in action in an animated piece, in most theaters, preceding every movie. This is not entirely whimsical. It's another example of preparing the public consciousness for coming events through an apparently fictional device.

4. Cloning and creation of artificial life-forms. The DIA-6 group identified five alien groups with which we have had experience. In this e-mail posted in Release 23, Anonymous identifies the five alien types, and a member of the group explains what the five types of aliens have in common. This will give the reader some idea of the incredible biotechnological capabilities of the Eben scientists.

> **VICTOR:** Regarding your question as to where some of the aliens come from, here is the complete list of the alien species the U.S. government knows of and has catalogued:
>
> 1) Ebens = Planet SERPO in Zeta Reticuli
> 2) Archquloids = Planet PONTEL near the Cygnus Arm
> 3) Quadloids = Planet OTTO in Zeta Reticuli (genetically engineered "praying mantis" and lizardlike creatures created by the Ebens)
> 4) Heplaloids = Planet DAMCO near the Cygnus Arm
> 5) Trantaloids = Planet SILUS in Zeta Reticuli
>
> **NOTES:** Both planets DAMCO and PONTEL are in the Milky Way near the Cygnus Arm. DAMCO is from a solar system that contains 11 planets and it is the (4th) planet out; its size is a little bigger than that of Earth. PONTEL comes from another solar system entirely with a sun just about the size of ours. PONTEL is the (5th) planet out and just a little smaller than ours. More to come.
>
> —ANONYMOUS

Here is an e-mail from another member of the DIA-6.

> There are (2) common links between each alien group and the Ebens. The first link is that the Ebens discovered each group, civilized them, and then CLONED their species with others. It is an extremely complicated subject, and is something I don't care to go into at this time with you. While we don't know all of the specifics, basically the Ebens used the DNA of each alien group to CREATE OTHER SPECIES OF ALIENS.

The second common link is the DNA. Each alien group has the SAME EXACT DNA. How that is possible, we don't know. Level 2 at the S-2 facility is where J-Rod and the other alien [Archquloid] lived. They have special containment facilities to house each alien.* Victor, one thing that has never been disclosed is the name given to the second alien, the one shot [the Archquloid]. The USG named that alien CBE-1, or CLONED BIOLOGICAL ENTITY-1. THIS HAS NEVER BEEN OFFICIALLY DISCLOSED TO THE PUBLIC; you have a "first" here.

The word Archquloid was coined by A51 [Area 51] scientists to classify each different alien race. We knew of (5), all given to us by the Ebens. We developed other names for each race especially for the Archquloid, who was the creature shot in the Gate 3 incident.† The Ebens cloned other races of aliens. As one of my other colleagues recently wrote and told you, it is really an incredibly complex story. But the Archquloid was a CLONED BIOLOGICAL ENTITY [CBE-1], created by the Ebens. It would take hundreds of hours and thousands of pages of written clarifications to explain it all, neither of which I am prepared to do. The QUADLOIDS were also genetically created by the Ebens. The Quadloids were cloned from (2) other species. So, as you can see, it really gets very complicated; hence my reluctance to delve into it too deeply.

5. Viewing/recording past events. The Yellow Book, presented to us by the Ebens as a gift upon their arrival in April 1964, displayed remarkable technology. Here is a complete description by Anonymous of the marvels of the Yellow Book.

(1) Reference the Yellow Book: It is an 8 x 11 inch object and approximately 2½ inches thick that is constructed of a clear, heavy, fiberglass-type

*Area S-2 is part of the sprawling Area 51 complex adjoining Groom Lake. J-Rod, or alternately J-ROD, was a genetically altered Eben, sent here for us to study. Anonymous says about him, "J-Rod was a creature created by the Ebens. He was intelligent, contained a brilliant mind, and was able to adapt quickly to our environment." J-Rod communicated telepathically with the Archquloid (see plate 17).

†The Archquloid killed a guard at Gate 3 while escaping from his compound. He was then shot and seriously wounded by another guard. He died a year later.

material. The border of the book is a bright yellow, hence the Yellow Book.

When you place the book close to your eyes, you will begin seeing words and images flashing before you. Depending on the particular language you are thinking, that particular language will appear. So far, the USG has identified 80 different languages.

Pictures also appear. The Yellow Book tells the story of the Ebens' lives, their exploration of the universe, their planet, their societal life, and other aspects, including the Ebens' longtime relationship with Earth. It tells of their first visit to Earth about 2,000 years ago. It displays Earth as it was in those days. It also shows an Eben who took the appearance of an Earth Human [Jesus Christ]. According to the Yellow Book, this Earth Human established religion on Earth [Christianity] and posted the first ALIEN AMBASSADOR on Earth.

The Yellow Book goes on and on and on and on . . . I've spent 12 hours a day for three (3) consecutive days and still NEVER reached the end. I don't think anyone knows how long it goes on or that there is any way to find the "end" of the book. There is NO known end to the Yellow Book. I understand the record is about 22 straight hours, which was done by the President's Scientific Adviser under the XXXXX administration. Also, there is NO known way to stop reading in one particular place, put down the Yellow Book, and then resume at the same place. Once you put down the Yellow Book and pick it up again later, the book starts from the beginning. Although the book can somehow determine the language of the person who is reading, it canNOT determine the uniqueness of that person. In other words, the "problem" with "reading" the Yellow Book is that one must start all over from the very beginning once you put it down. If it took you 12 hours to reach p. 564 [events] and then you put it down, you would have to start all over at the very first words and images displayed during your first read, and only once you passed the 12-hour mark would NEW information you had not seen in the previous sitting appear.

As I wrote earlier, the Yellow Book goes back to around 2,000 years. However, I have NOT viewed the entire Yellow Book and don't think anyone else has, either. It may have some views/images/history that go back even farther than 2,000 years B.C. [?].

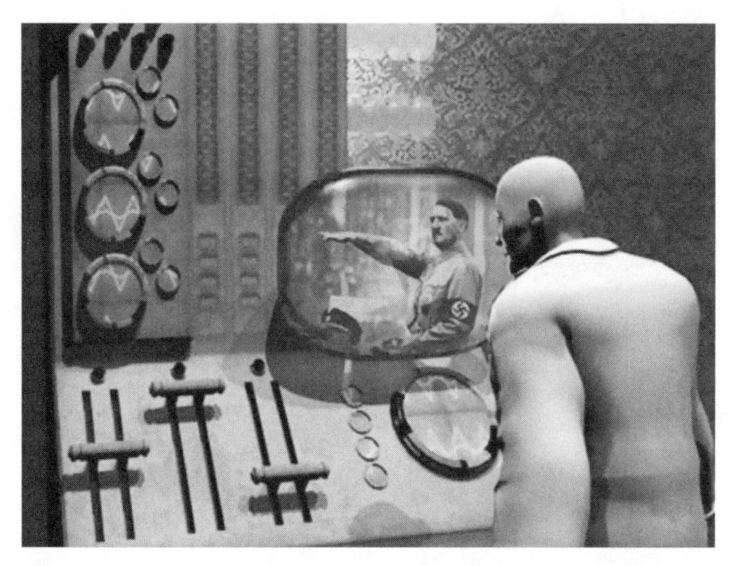

The Chronovisor, a device for seeing past events developed by two Italian priests (see plate 27), apparently duplicates the Yellow Book technology. Image copyright Lee Krystek.

The Yellow Book has the following characteristics:

- It can read your mind to determine which language you think in, and then display text in that language. This means that once it is within your aura, it can detect and access your mental field, and can then read your thoughts and match your language from a vast database.
- It displays actual images from the past, going back as far as two-thousand years. That means that the Ebens have the ability to actually return to the distant past, and to record images when there. The storage and display of those images in the book itself is not especially amazing, since we are rapidly approaching that capability ourselves, although in 1964, that would have appeared to be miraculous. An alternative explanation is the possibility that the Ebens can access what we call the "Akashic records," and can somehow record the events within that access. This is actually a more likely explanation, since Edgar Cayce was able to access the Akashic records selectively. So if the Ebens can do that, they can choose to

see and record only the events from our past that are significant.

- It selects only meaningful events of the past; for example, the advent of Jesus.

6. Interstellar communications. Anonymous says,

> Los Alamos "Project Gleam" is a highly classified project that deals with direct communication with "The Visitors." New communication technology dealing with multifrequency sending units. Units direct multiple frequencies in a particular direction. High speed sending system allows the beam to be propelled at an enormous speed. Not too much more known about it. Los Alamos and several contractors, including EG&G, BDM, Motorola, Risburn Corporation, and Sandia, are [all] involved with this project. Facility built at Site 40, Nevada Test Site.
>
> One rumor (only rumor from my source) is that The Visitors provided us with this technology. It enables us to communicate with The Visitors in a speedier way than in the past. Part of this program involves the use of chemical lasers pushing the communication beam.
>
> As it was explained to me (in layman's terms), several frequencies are put together on a beam and propelled toward a target or receiver. The receiver then boosts the energy and re-sends the signal to another point (relay?). Somehow, the chemical laser pushes the beam, thus propelling it faster than normal.

7. Translation devices. The Ebens brought translating devices with them at the first landing in 1964. Anonymous describes them in his narrative.

> The Ebens had a crude translator device. It appeared to be some sort of microphone with a read-out screen . . . The senior U.S. official was given one of the devices and the Eben kept the other one. The officials spoke into the device and the screen showed a printed form of the voice message, both in Eben and English.

Evidently, the devices communicated with each other wirelessly. This seemed like amazing technology at that time, but we have since achieved and surpassed this capability now. Dragon software now instantly translates voice input to written form, which can be displayed on a computer or iPad screen, and sent wirelessly to another computer or iPad. Also, Google has now made available instant language translation from/to most common world languages, as part of its search engine. As with many of our other technological leaps, this capability could have resulted from what we learned from the Ebens.

8. Sound weapons. As reported in the Team Commander's diary and previously discussed, the Eben guide accompanying the team in their exploration of the planet had some sort of weapon that projected a sound wave and repelled a dangerous creature encountered by the team in their travels. In this case, the Commander described the creature as "like an armadillo." There has been extensive R&D in this country and abroad in the use of acoustic weapons. We have already developed the capability to use sound in certain frequencies to disperse crowds, and we can already use powerful sound waves to disorient and to render neurological damage. Development of an acoustic weapon that kills is clearly within our reach.

Long-range acoustical device, developed by the Navy

18

THE MOVIE

I really found my faith when I learned that the government was opposed to the film. If NASA took the time to write me a twenty-page letter, then I knew there must be something happening.

STEVEN SPIELBERG

Spielberg tells me about the film. It's about UFOs, he says, but it is not science fiction. He calls it "science fact."

BOB BALABAN, ACTOR IN
CLOSE ENCOUNTERS OF THE THIRD KIND

THE SPIELBERG MAGIC

The Serpo story took on a new dimension when people started to realize that the Steven Spielberg movie *Close Encounters of the Third Kind,* released in 1977, may have been a fictionalized version of the actual event. The story, revealed twenty-five years after the movie was released, came clothed with a convincing aura of believability by virtue of the retrospective appreciation of the link to the movie, which had become a landmark film on its own merits. That appreciation, and the fact that *Close Encounters* did become an enduring classic, is a testament to the startling neorealistic way that Spielberg tells fantastic stories. In

Spielberg movies, the unbelievable becomes very believable. Because of the way that fact and fiction are seamlessly woven together, the audience leaves the theater convinced, at a deep level, that what they just saw actually happened. That technique explains the impact and success of his first major movie, *Jaws*. Amazingly, in that film, he was able to achieve that effect even though he had to work with a clumsy mechanical shark. So, when the Serpo revelations appeared on the Internet in November 2005, they fell on the already prepared public consciousness due to the power and lingering effects of *Close Encounters,* which had, by then, achieved a permanent place worldwide in the mass culture.

PUBLIC ACCLIMATION

Many in UFO circles have come to believe that Steven Spielberg has an inside track on top secret military information, and that he doesn't just "make this stuff up." There is some evidence that he has been chosen by the Pentagon, because of his neorealistic filmmaking skills and his popularity, to participate in a plan for the slow release of classified information, the so-called "public acclimation program." Belief in the likelihood of that connection is supported by his choices of subject matter. Several of his films can easily be related to possible government/military agendas for the dissemination of secret information in a way calculated to win public sympathy. For example, in *E.T.: The Extraterrestrial*, released in 1982, we respond with affection to a rather grotesque alien creature, who becomes lovable by virtue of his "heart light."

But in the case of *Close Encounters of the Third Kind,* the motivations are less clear. In fact, they are downright mysterious. One must ask the question: If this movie was part of the public acclimation program, why was it released while this top secret DIA/Air Force operation was still in progress? Spielberg began filming *Close Encounters of the Third Kind* in 1976. But the Serpo astronauts did not return to Earth until 1978. If the goal was to condition the public to an eventual revelation of the interstellar exchange program, that information would not have been released until the program was successfully completed, because it could have ended in disaster if the astronauts never returned to Earth. Furthermore, if it is suggested that perhaps it was Dr. J. Allen Hynek who gave Spielberg all the details of the Serpo operation, that likelihood can be easily dismissed. Hynek was the main UFO/ET technical consultant for the film and therefore was well-placed to inform Spielberg.

However, despite Hynek's long association with secret government UFO activity, nothing he ever wrote or said gives any hint that he knew anything at all about Project Serpo. Furthermore, there is every reason to believe that Hynek's participation was largely ceremonial. He was brought into the movie partially because he had conceived the "close

Dr. J. Allen Hynek

encounters" scale in his book *The UFO Experience,* published in 1972, and this was a way of compensating him for the rights to the title. He himself said that his assistance with the film was rather superficial, and that he was mainly interested in learning more about how movies are made. As perhaps further compensation, he was given a cameo role in the last scene of the movie.

THE SCRIPT

By 1974, Spielberg had become identified as one of the rising directorial stars in Hollywood. He was one of the prominent members of the "movie brats" generation. They, especially George Lucas, Martin Scorsese, Francis Ford Coppola, and Brian De Palma, were the talented young directors of the 1970s who were clearly destined for great things. Spielberg's work on *Sugarland Express,* and the huge success of *Jaws* got him a longterm contract with Universal and the privilege of choosing his own projects. He was also free to work with other studios. He was given a semi-private bungalow on the Universal lot and was meeting the right people. He became friendly with Michael Phillips who, together with his wife Julia, were codirectors of Phillips Productions. They were riding high, at that time, having produced *The Sting,* which collected a basket of Academy Awards including Best Picture of 1973. Julia Phillips considered Spielberg one of the most talented people in Hollywood, and she pushed for a project with him. He told her he wanted to make a film about UFOs. She was all for it and began to assemble the pieces to make it happen. Phillips was in the middle of the production of *Taxi Driver* at that time, working with screenwriter Paul Schrader and David Begelman, president of Columbia Pictures. Schrader had written a script he called *Kingdom Come* about a dedicated civilian UFO investigator who gets in trouble with the Air Force. In the end, he decides to leave the planet and go off with his alien friends. Sci-fi was an anomaly for Schrader, who was better known for his work with harsh realism. Phillips brought Schrader's script to Spielberg, who said it was a good starting point, but would need an extensive rewrite. Phillips bought the script and

Spielberg changed the name to *Close Encounters of the Third Kind.* She then easily persuaded Begelman, who immediately liked the concept, to have Columbia finance and distribute the movie.

Hollywood producer Julia Phillips (1944–2002)

According to Julia Phillips, in her book *You'll Never Eat Lunch in This Town Again,* Spielberg sat down and rewrote the entire script in one weekend in 1975 at the Sherry Netherland Hotel in New York City, so that almost none of Schrader's work remained in the finished product, and Schrader ended up with no screenwriting credit. For Spielberg to have written such a complex screenplay so fast, in a hotel, without doing the research this script obviously required, would indicate that the story was given to him in advance. Since we now know that all the facts of the screenplay was in accord with the actual events of Project Serpo, that is the only plausible conclusion. The script became basically locked in place after that, with only minor modifications approved by Spielberg. That was a very unusual process in Hollywood, where screenplays are generally written by screenwriters, not directors, frequently take weeks or months to write, and normally go through many changes, based on the input of the producers and studio execs. Not this one. With the backing of Julia Phillips and Begelman, nobody but Spielberg, or someone he recruited, could change one word.

A SECRET SET

And then there was the secrecy. The *Close Encounters* set near Mobile, Alabama, took on the aspect of a top secret military facility. According to John Baxter in *Steven Spielberg: The Unauthorized Biography:*

> Mobile became the most hermetic of closed sets. Cast and crew worked, ate, sometimes even slept inside the stifling hangar, which 150 tons of air-conditioning equipment, enough for thirty large houses, did little to make habitable. . . . Nobody entered without a name badge. Even Spielberg himself, who lived in a Winnebago parked outside, was briefly barred when he forgot his. Scripts were numbered and distributed on a need-to-know basis. Most actors got only their own lines.

It's understandable and not unusual for directors to not want plot information leaked in advance of the planned publicity blitz. But they also want to build advance interest in the movie, so small, controlled leaks are desirable. In this case, the security was so severe that, clearly, something else was operating. It would be fair to speculate that Spielberg perhaps knew that he was dealing with the release of highly classified information and didn't want to risk having the government shut him down on the basis of the compromise of national security.

THE MOVIE COMPARED TO THE EVENT

If the Serpo information was not given to him by the military as part of public acclimation, and if he didn't get it from Hynek, then from what source did Spielberg obtain all the correct details of a program that was so super-secret and compartmentalized that only a very few top intelligence and Air Force operatives knew anything about it? Very likely, President Johnson himself was kept out of the loop. Johnson was not

Steven Spielberg filming Close Encounters of the Third Kind *in 1976*

known to have been among the sixteen in the diplomatic greeting party at the first landing of the Eben craft in April 1964 near Holloman Air Force Base. In chapter 8, we discussed the fact that the aliens decided not to take the astronauts at that time, even though they were ready and waiting in a bus at the landing site, but rather to just take the ten dead bodies of their Eben compatriots, and that landing was not depicted by Spielberg, But he had all the basic facts right about the second landing in July 1965 at the Nevada Test Site. The movie agrees completely with all the details revealed by Anonymous and sent to the Serpo website by Victor Martinez. The details that Spielberg changed were mostly for dramatic effect. He knew that the visit was prearranged. He knew that this was not the official diplomatic greeting, which had taken place in 1964. He knew that it took place in summer since, in the film, everyone was in shirtsleeves in cold northern Wyoming. He knew that the Ebens communicated in a musical language, hence the five-tone greeting, and the musical "conversation" after the mother ship landed. He knew that the astronauts arrived on a bus. In the film, the astronauts are shown boarding a Greyhound bus. He also knew that sign language was used by the visitors to communicate with Earthlings. This was an esoteric piece of information that was only revealed twenty-five years later in a single sentence in the sixth posting by Anonymous on the Internet. He knew that ten men and two women were selected for the trip, or so he believed from credible evidence. He knew that they were paramilitary and that they were highly trained and disciplined, as is evident from their martial demeanor as they boarded the spacecraft. He knew that they needed super-strength sunglasses to protect their eyes from the punishing rays of the two suns beaming down on Serpo. They were already wearing the glasses as they boarded the alien spaceship. He knew that one Eben remained here on Earth. He knew that the astronauts took tons of supplies with them. In the movie, the supply truck was disguised as a Piggly-Wiggly van. He knew that the spaceship was massive, as it had to be to accommodate over forty-five tons of astronaut equipment and supplies. And he knew that the president wasn't present. And then there were other subtle intimations that Spielberg had inside information; the minister at the final church service before departure referred to the astronauts as "pilgrims," and the astronauts attending the service appeared anxious, almost desperate. This implied that they knew that they would be gone from Earth for a long time. The plan was that they remain on Serpo for ten years. That means that Spielberg must have known that, and had instructed the actors to display those emotions.

It's not hard to understand why Spielberg changed the landing site from the barren landscape of the Nevada Test Site, where the July 1965 landing actually took place, to the dramatic Devil's Tower in the Black Hills National Forest in Wyoming. This change allowed him to insert the entire first act of the script, wherein Roy Neary became obsessed with the implanted image of the Devil's Tower. Neary's obsession was due to a telepathic implant of that image in his mind during a "close encounter of the second kind" in his truck. This is a sophisticated concept that could only have been understood by a very knowledgeable insider. That phenomenon occurred with many contactees, going all the way back to George Adamski. And it is precisely what happened to Miriam Delicado, author of *Blue Star: Fulfilling Prophecy,* published in 2007, who had a close encounter of the third kind in 1988 on a lonely road in remote British Columbia, and subsequently became obsessed with Ship Rock in northern New Mexico before she even knew it existed. But Spielberg would have to have read a lot of books to have

The Devil's Tower in Wyoming

ferreted out that piece of information from the prodigious amount of UFO contact literature, and then grafted it onto the Serpo story. That was unlikely. Someone high up in the intelligence apparatus must have conveyed that idea to him.

But that part of the story has no basis in fact. Anonymous makes no mention of anyone wanting desperately to board the alien ship, and being allowed to do so. It is actually a weakness in the plot, since it makes no sense that Neary would just abandon the human race, including his family, to impulsively go to a distant planet and live with aliens. He had no real motivation to do that, unless we are to understand that he was really an alien in human form. That is actually implied in the film because it was the Ebens who had initiated his obsession, and they affectionately guided him onboard the craft at the end. Oddly enough, though, in reality, after thirteen years on Serpo, two of the astronauts did decide to remain there and did not return to Earth. In addition to Neary's obsession, Spielberg also grafted another advanced concept onto the story—time travel. The World War II Navy planes of Flight 19 that disappeared in 1945 off Fort Lauderdale, Florida, were found in the

Flight 19 disappeared in the Bermuda Triangle.

Mexican Sonoran desert intact and operational in "present day" in the film, which was 1975. It appears that the planes had gone through a time warp to the future, and the pilots were abducted by the aliens. The pilots were then released from the mother ship, along with other abductees, in the last scene of the movie. Spielberg later addressed time travel, another of his major interests, more thoroughly in *Back to the Future.*

A SCI-FACT FILM

It is very evident from all this that Spielberg must have had a "Deep Throat" giving him detailed information about Project Serpo. And, because of the timing, it is also clear that this was not part of an official public acclimation plan intended to condition the mass consciousness, but rather a deliberate leak by a rogue insider who wanted the public to know what was going on. It is now known that there are many in high military ranks who are in this category. But this particular leak probably originated in the Defense Intelligence Agency, because only they knew all these details about the program, and Anonymous, who ultimately released the story on the Internet in 2005, was a retired DIA official. Since Anonymous appears to have been dedicated to getting this information out to the public, quite possibly he was the one who gave the story to Spielberg thirty years earlier! He would, at that time, have been actively involved in the project. In any case, most likely, an idealistic, disaffected, top DIA official came to Spielberg, because of his success with *Jaws*, and offered him the story of a lifetime. The young Spielberg, already a science-fiction film enthusiast, seized the opportunity and made a sci-fact film dressed in science-fiction apparel.

Close Encounters was Spielberg's first venture into the world of government UFO/ET secrets, and it was definitely not officially sanctioned at any level. Both the Air Force and NASA refused to cooperate in the making of the movie. It grossed almost $300 million internationally, and it saved Columbia Pictures from insolvency. Ray Bradbury pronounced it the greatest science-fiction film ever made. It was nominated for nine Oscars at the fiftieth Academy Awards in 1978. In 2007, *Close Encounters* was declared "culturally, historically, or aesthetically significant" by the U.S. Library of Congress and was selected for preservation in the National Film Registry. It was only after the spectacular success of this movie that top military and intelligence officials evidently took notice of Spielberg, and realized that he was the perfect choice to make future films for the public acclimation program.

Spielberg might have learned from his DIA contact that it was President Kennedy who had issued the edict for Project Crystal Knight, and that he was assassinated only five months before the first landing. That much is known. But he might have also been informed that Kennedy, who detested government secrecy, had planned to make the story public, and possibly have the landing televised. And Spielberg, who was seventeen when Kennedy was killed and was probably crushed by the assassination, as we all were, might have then made up his mind that he would carry out, if only as a staged version, what our dead, beloved president had intended. That, of course, is speculation. Regardless, thanks to Steven Spielberg, we got to see it after all.

APPENDIX 1

TEAM TRAINING CURRICULUM

This training program was sent to Anonymous by list member Gene Loscowski (see the introduction) in the interest of adding more specificity to the released information. Anonymous then reproduced the list and submitted it as part of Posting 10.* Obviously, Loscowski was involved with Project Serpo, but it is not known whether or not he was a DIA official.

1. Introduction to Space Exploration (taught by NASA personnel)
2. Astronomy, identification of stars, use of telescopes and general astrophysics
3. Eben anthropology (information received from Ebe1)
4. Eben history (basic information received from Ebe1)
5. U.S. Army field medical training (trauma care). This was given to the nonmedical personnel on the team.
6. High-altitude training—parachute and weightless/zero oxygen environment training [probably given at Tyndall AFB in Florida]
7. Survival, escape, and evasion training
8. Basic weapons and explosive training (six pounds of C-4 [Composition-4] was taken)

*The "Posting" designations by Anonymous ended in Posting Eighteen. From that point on, the term "Release" was used.

9. Psychological operations training and anti-interrogation preparation
10. Small unit tactical training (mini-week U.S. Army Ranger course)
11. Intelligence-gathering course
12. Space geology—collection methods and use of specialized geological equipment
13. Physical stress training
14. Methods to cope with confinement and isolation
15. Nutrition course
16. Equipment use training
17. Individual specialty training
18. Basic biology
19. Other training that is still considered extremely highly classified even after 40 years [1965–2005].

APPENDIX 2

SUPPLIES AND EQUIPMENT

This list was submitted to the e-mail network on April 3, 2006, by Anonymous as "Posting Eighteen," with the following message: "Victor: The following information pertains to the equipment our Team took to Planet SERPO. All of the military equipment was real . . . at least back then." By "real" Anonymous evidently means that it was operational. Evidently, this information was included in the debriefing document.

1) MUSIC; The Team members took the following types of music:*

Elvis Presley
Buddy Holly
Ricky Nelson
The Kingston Trio
Brenda Lee
The Beach Boys
Bob Dylan
Peter, Paul & Mary
The Beatles
Loretta Lynn
Simon & Garfunkel
The Hollies

*Audio cassette recording was introduced in 1963, so it is likely that all of these recordings were on cassettes, although some may have been on reel-to-reel tapes.

Chubby Checker
Bing Crosby
Dinah Shore
Vera Lynn
Tommy Dorsey
Ted Lewis
Ethel Merman
The Everly Brothers
Lesley Gore
Marlene Dietrich
The Platters
Doris Day
Connie Francis
The Shirelles
Frank Sinatra
Dean Martin
Perry Como
Guy Lombardo
Glenn Miller
Rosemary Clooney
Al Jolson
Christmas Music
U.S. Patriotic Music
Classical Music:
- Mozart
- Handel
- Bach
- Schubert
- Mendelssohn
- Rossini
- Strauss
- Beethoven
- Brahms
- Chopin
- Tchaikovsky
- Vivaldi

Indian Chanting Music
Tibetan Chants
African Chants [these last three for the intended benefit of the Eben hosts]*

2) CLOTHING; The team members took the following clothing:

24 pairs of specialized flight suits
112 pairs of underwear (pants/shirts)
220 pairs of socks
18 hats, including jungle-style and regular ball caps
50 different types of footwear
military clothing, load bearing belts and harnesses
military backpacks
30 pairs of civilian casual pants
shorts
sleeveless shirts
15 pairs of athletic shoes
100 pairs of athletic socks
eight (8) athletic supports
24 pairs of thermal underwear
24 pairs of thermal socks
six (6) pairs of cold weather boots
military-style hot weather clothing
60 pairs of gloves military work-style
10 containers of military-style sanitary gloves
six (6) pairs of cold weather gloves
10 laundry bags
disposal surgical gloves
military-style warm weather jackets
military-style cold weather jackets
civilian-style warm and cold weather jackets
10 pairs of warm weather sandals

*Since we knew this much about Eben musical interests, it validates our previous conclusion that we had detailed information about their civilization prior to the spaceship departure in 1965.

24 military safety helmets
24 military-style flight helmets
1,000 yards of fabric for the repair and making of clothes

3) MEDICAL EQUIPMENT; The team members took the following medical equipment:

Portable X-ray machine
100 prepacked medical kits for advance[d] trauma care (military-style battlefield medical kits)
examination scopes for the stomach, bladder, and rectum
eye examination equipment
120 prepacked surgical kits (military-style)
120 prepacked military field medicine kits (containing various medicines)
30 military-style field medical sanitation kits
75 water-testing kits (military-style)
50 water-testing kits (civilian)
75 FAST kits
1,200 food-testing kits (military-style)
500 pieces of miscellaneous surgical tools
5,000 packages of insect repellant (military-style)
250 medical intravenous kits, with fluids
16 prepacked medical testing kits (military-style)
50 prepacked medical testing kits (civilian)
five (5) military Medical Portable Hospital Tents with base
two (2) Military Medical Portable Deployment Kits
18 Military Medical Blood testing kits
three (3) portable military chemistry testing stations
two (2) Advanced Biological Testing Kits (civilian version)
15 Military Radiation Treatment Kits
1,000 pounds of miscellaneous medical equipment

4) TESTING EQUIPMENT; The team members took the following testing equipment:

100 pieces of geological testing items

2 military soil testing stations
two (2) chemistry testing stations (civilian)
six (6) radiation testing meters
two (2) military radiation testing stations
two (2) biological testing stations (civilian)
two (2) 100-cc tractors
four (4) 100-cc digging-tool tractors
10 pre-packed military Soil Testing Kits
16 astronomical telescopes
two (2) Military Star Stations
four (4) military power generators (1–10,000 watts)
four (4) civilian power generators
experimental solar collecting equipment (military)
50 portable two-way radios with FM frequencies
six (6) military combat radio platform kits
50 prepacked military radio repair kits
1,000 different-frequency tubes
30 prepacked military-style electrical testing and repair kits
three (3) solar-testing stations (military)
one (1) experimental solar testing station
10 solar collection panels with collector containers
10 air sample collection kits (military)
five (5) air sample collection kits (civilian)
six (6) diamond drills
10 military special access kits
1,000 pounds of C-4 explosives with 500 blasting caps
detonating cord
time fuse
Military shape charges
one (1) Nuclear Detonating Kit

5) MISCELLANEOUS EQUIPMENT and ITEMS; The team members took the following miscellaneous equipment/items:
100 military blankets
100 military sheets

24 prepacked military combat deployment kits
80 prepacked military combat tent kits
four (4) military mobile kitchen deployable kits
six (6) military survival stations warm weather
six (6) military-style survival stations cold weather
2 military weather stations combat style
50 military weather balloons
24 military handguns
24 military rifles (M16s)
six (6) M66 weapons
two (2) M40 grenade launchers
two (2) military 60 mm motor tubes (30 rounds)
100 military air-burst flares
5,000 rounds of .223 ammunition (for M16 assault rifles)
500 rounds of .45 ammunition
60 M40 rounds
15 Freon dispersal containers
15 compressed air dispersal containers
20 tanks of oxygen gas
20 tanks of nitrogen gas
20 tanks of miscellaneous gases for cutting equipment and testing
75 military-style sleeping bags
60 military-style pillows
55 military-style sleeping platforms
six (6) prepacked military deployment combat field living platforms
250 different style padlocks
6,000 feet of different types of rope
24 repelling [*sic*] kits
10 seismic deep-hole drills
1,000 gallons of fuel
four (4) military-style phonographs
10 Military cassette players
10 reel-to-reel tape players
60 belts
10 military sound collection equipment kits

25 military Intelligence Collection Kits
1,000 other miscellaneous items

6) VEHICLES; The team members took the following vehicles:

10 military-style combat motorcycles
three (3) military M151 Jeeps
three (3) military trailers
10 Military repair kits for Jeeps
10 Military repair kits for the motorcycles
One (1) Military lawn mower*
1,500 gallons of fuel for all of the above items

7) FOOD; The team members took the following food items:

C-Rations
25 prepacked containers
100 prepacked containers of freeze-dried food items
100 cases of various canned food items
seven (7) years worth of vitamins
100 containers of energy bars/snack items
1,000 gallons of water
150 military survival food kits
16 boxes of various alcoholic wines
150 cases of drinking fluids
Chewing gum, lifesaver candy and various other miscellaneous food items

8) MISCELLANEOUS ITEMS; The Team Members took 2,000 pounds of various other items.

*Evidently they thought they could use the lawn mower motor for other purposes.

APPENDIX 3

SERPO STATISTICS

Anonymous submitted these statistics as "Posting Three" on November 7, 2005, with the following introductory message: "Statistics on the Eben planet was [*sic*] collected by our team. Here is the pertinent data for your UFO thread list:" Evidently, these data were gathered by team members while on Serpo.

Diameter:	7,218 miles
Mass:	5.06×10^{24}
Distance from Sun 1:	96.5 million miles
Distance from Sun 2:	91.4 million miles
Moons:	2
Surface gravity:	9.60m/s
Rotation periods:	43 hours
Orbit:	865 days
Tilt:	43 degrees
Temperature:	Min: 43°/Max: 126°
Distance from Earth:	38.43 light-years
Planet named by team:	SERPO

Nearest planet:	Named: OTTO
Distance:	88 million miles (colonized by Ebens with research base, but no natural inhabitants on planet)
Number of planets in Eben Solar System [Zeta Reticuli 2]:	Six
Nearest inhabited planet:	Named: SILUS
Distance:	434 million miles (made up of creatures of various types, but no natural inhabitants on the planet)

APPENDIX 4

SERPO PLANETARY MOTION AND TIME MEASUREMENT COMMENTS

The disclosures by Anonymous stimulated a lively discussion. Many comments sent in to the website addressed believed scientific discrepancies and anomalies. Others sought to solve the time dilemma that caused the team to come home three years late. All the comments are reproduced here in the interests of perhaps filling in missing pieces and of adding more depth to what Anonymous has revealed. I have tried to present them faithfully verbatim, but in some cases minor corrections were necessary.

> After reading Dr. Sagan's remarks on the Serpo project, which is about sixty jam-packed pages of calculations, I found one paragraph which states that in order to use Kepler's law in the case of Planet Serpo, one had to vary the exact gravitational pull placed on Serpo by the two suns. Serpo did not have large planets, like Jupiter and Saturn to affect the gravitational pull, as the Earth does. Serpo's gravitational pull was different than anything Dr. Sagan had ever seen before. There are numerous figures and calculations to support this. I will forward them at a later date. Have your list stay tuned . . .
>
> In response to the intensive, ongoing debate recorded in the Comments section of this site, I would like to endorse this clarification from a prominent physicist on the list: *The senior theory (essentially proven) is*

Newton's inverse square law for gravity. For a simple system, Kepler's laws are a fallout of what a planetary solution looks like (for the simple case of a single planet circling a massive sun). For a complex situation (like a planet interacting with two suns, several planets, or whatever), you have to go back to Newton's law and solve a many-body problem, which takes a computer. In this case Kepler's laws are only an approximation, since they hold only for the simplest case. —Anonymous

There are conflicts with other data. Rick Doty, who witnessed the Ebe2 interview* says Zeta 1 had eleven planets. . . . Zeta 2 was well beyond the orbit of the eleventh planet. Zeta 1 and 2 are not close binaries. They are spread very far apart. Astronomical observations back that up. Rotation period according to Ebe2 was thirty-eight hours. Temperatures: 65°F–90°F. Orbital tilt was 54 degrees. All this from the Ebe2 interview in the book [see footnote]. Epsilon Eridani is a K2-type star. Age estimates put it at half a billion to a billion years old. Not long enough for even amoebas to develop. ANONYMOUS has got lots of things right, but other things wrong, or it's mixed up, which makes me suspect.

—Comment November 7, 2005

Nothing really new here except added information on the makeup of the planet. As for the laws of physics, your Anonymous is twisting the facts. The orbits of those two stars, Zeta 1 and 2, are an observational fact, not some law of physics, although all galaxies, stars, and planets operate under those laws, from what's been observed. The two stars are spread wide apart by 350 billion miles, called a wide binary. And Kepler's law was applied to the planets of Zeta 2 and it works pretty [well] even for elliptical orbits. Puts Ebe1's home planet . . . right where it should be, see table. The game I think is to convince everyone that, yes, Anonymous does have some good information. But beware: good information from what I see is being sprinkled with bad information or disinfo. For Zeta 1 and 2, see link (http://www.solstation.com/stars2/zeta-ret.htm). For Zeta 2 planets, see table. Ebe1 came from the fourth

*Reported in his book *Exempt from Disclosure*. See chapter 16.

planet around Zeta 2. See the AU [Astronomical Units] place planet 4 is sitting around Zeta 2 or 1.12 AU. Very nice place to live. . . . Notice that Zeta-4 orbital period is 432 days, but yet it's farther out than SERPO was in that Anonymous report, which gave a ridiculous 865.

PLANETS OF THE ZETA 2 RETICULUM SYSTEM

PLANET	SEMI MAJOR AXIS	PERIOD (DAYS)	PERIOD (YEARS)
Reticulum 1	0.14 (AU)	8.9	0.052
Reticulum 2	0.28	54.0	0.1481
Reticulum 3	0.56	152.9	0.4196
Reticulum 4*	1.12	432.6	1.12

*This is Serpo

So one Reticulum 4 year is equal to roughly 1.12 Earth years, or 432 days. And it is in roughly the same position in Zeta 2 Reticulum's "life-zone" as the Earth is in the sun's. Zeta 2 Reticulum is a G1V spectral class star, the sun is a G2V.

—Comment November 9, 2005

If the "Away Team" couldn't make their measuring devices work properly, then why even quote numbers like these? Is a mile on Earth a mile on SERPO? Recall that lengths are now referred to as the wavelength of light. Is the light wavelength to be trusted? Does red light generated by excited neon atoms, which has a wavelength of 6328 angstroms on Earth, have some other value in SERPIAN length measurement units? Is there neon on SERPO? (Are their elements the same [as ours]?) Some have speculated that there could be different values for Planck's Constant, ***c***, etc., in different universes. If SERPO is in another universe, all bets are off as to whether or not we could make sense of what is "normal" to them. (Talk about "out of the box" thinking!) Some of you may recognize this as the wavelength of red HeNe laser light. If an HeNe laser were built on SERPO and the laser light directed toward Earth, would the wavelength when received on Earth be 6328 Å? If so, then there should be a simple ratio between SERPIAN length units and ours.

If not, then it would be much more difficult or perhaps impossible to derive SERPO physics from our own. Moons: SERPO has two. Complicated tides? Distances from planet? Relative sizes? Periods? Day: 43 hours. Whose hours? Makes no sense to quote this as hours unless it is intended to mean our hours, which are related to our seconds (3,600 per hour) and each second is a number of oscillations of the atomic clock. If atoms "run differently" on SERPO, then not only will wavelengths be different (see above) but also frequencies (time durations of oscillation). Which type of day? Does this refer to rotation of the planet relative to one of the nearby stars or relative to the distant stars? (We measure "sun days" and "distant star" days; they differ slightly because the Earth rotates around the sun.) ASSUME this "day" is 43 of our hours, so there are 154,800 of OUR seconds in THEIR day as compared to 86,400 of OUR seconds in our day (approximations used liberally!) Year: 865 days. Does this mean our years? Probably not! For someone on SERPO, it would be natural to measure a complete rotation around a sun (relative to distant stars) in terms of the rotations of the planet. If we assume this means THEIR days of 43 hours, then their year is (all numbers approximate) 1.3×10^8 of OUR seconds, whereas our year is 3.1×10^7 of OUR seconds. Previously, the Kepler's law was applied under the assumption that the mass of the sun about which the planet rotates was the same as our sun. However, I, probably incorrectly, used 865 days as 865 of "our" days, which would be 2.37 of our years. I should have accounted for the 43-hour day. A day this long means that the rotation period is actually $865/365 \times 43/24 = 4.2$ of our years. For the Earth, Kepler's rule can be written as follows, using years and AU as the units of measurement (with 1 AU = radius of Earth from sun) $(1)^3/(1)^2 = 1\ (AU)^3/(year)^2$. For a planet rotating around a sun with the mass of our sun in 4.2 years, Kepler's rule is written as $r^3/(4.2)^2 = 1$, which leads to a radius of 2.6 AU or 240 million miles, noticeably larger than the 164 million I calculated before.

One problem with this sun data: if strictly interpreted, as the distance from one sun is ALWAYS 91.4 million [miles] and the distance from the other is ALWAYS 96.5 [million miles] then we have an "impossibility"... or at least I can't conceive of an orbit that keeps the distance from one constant at one value and at the same time keeps the distance from the other also at one value. I could conceive of an orbit that keeps the distance from one constant while the distance to the other varies with a periodicity that depends upon the period of rotation of the suns around one another and the rotation period (SERPO year) of the planet. If interpreted as average distances from the suns as SERPO travels in an elliptical (or more complicated) orbit, then there may be some way to explain this. Of course, those suns would be orbiting one another at some distance apart, so if SERPO did have an orbit that encompassed both suns, then it must orbit the center of mass of the suns (unless there is some complex orbit, as I suggested previously, like a figure eight or a warped ellipse). Need more info on the suns, orbit, etc.

—Comment, November 11, 2005

Neat! Of course an alternative is that the Suns are somewhat less massive than our Sol < (which would shove the mass ratio of Theirs/Sol 1 in place of the normalized "1.0" (with the Earth Orbital Distance, and the Earth Orbital Period, the respective Length and Time units), which would mean that larger Periods would prevail at similar Central Star-to-Planet Distance, or less distance result at similar Orbital Period, compared to the Earth-Sol system. If they were in fact OLDER (relative time over expected life span of such and such a mass star) than Sol, then they may already have begun Helium Burning and a climb in Brightness, which could allow a bit lower mass stars than Sol to shine as brightly. Now novel several-body configurations have been found occasionally over the last century, but most effort has gone into the restricted 3-Body problem IN A PLANE. I can imagine a possibility with two similar stars—there might be a chance of stable manifold for a small planet in a "slow wobbling orbit" around the center of mass of the two suns, approximately with "normal" axis the instantaneous line between the two stars. I write the quotes around "normal" because I expect the orbit to reside NOT in a Plane but RATHER in a torus centered on the center of mass of the two Stars.

However, the squeaker of a coincidence for the surface gravitational acceleration value (9.97 and 9.96, compared to engineering assumed

Earth normal of 9.806 m/s^2) you show below suggests that some of the "data" has been cooked in a whimsical way.

If Serpo is slightly similar in size than Earth but has a higher bulk mass density, then I certainly see a possibility of less of the lighter mantle elements O, K, Mg, S, Si than with Earth, certainly less water of hydration in upper mantle rock, and maybe less crustal pore space.

—Comment, November 11, 2005

I'm a military scientist. Here are my observations regarding the SERPO information: The question of whether the "laws of physics" apply differently from galaxy to galaxy dependent upon their spin is irrelevant to any discussion of Reticuli twins, in that they are both part of the same Milky Way galaxy as is our Sun. Moreover, any distortion in our observations of OTHER galaxies which MIGHT result from the spin of our own, would apply across the board, and thus in essence negate itself, since there would be no way to logically compare it to similar distortional effects which would affect observers in other galaxies in their efforts to observe us, presuming that both observations were initiated at the exact same moment of universal time. In other words, if a third observer in some other dimension were able to compare 1 unit of time (say, a single oscillation of one atom of cesium) as observed on planet Earth, with 1 unit of time equally measured on the planet X in the faraway galaxy of XXXXX, that observer might conclude (after removing all other variables) that the two units were different from his perspective, even though both observers in both galaxies might observe and report what they thought were the same results.

A cesium clock ticks off one second on planet Earth. A twin cesium clock placed in orbit and traveling at orbital velocity also ticks off one second. Observers at each location agree that one second has passed, as confirmed by the readout on their instrumentation. It is only when the two clocks are compared that it becomes obvious that there has been some distortion; yet in attempting to compare them it becomes equally obvious that the same observer cannot ever observe both at EXACTLY the same time (with a bow to Heisenberg). Thus, differences arise NOT from the observation, but from the perspective used to interpret it.

The basic theorem: Only from an observation platform set in ANOTHER universe would an observer be able to detect and have some basis of accurately comparing distortions being produced locally in THIS universe by differences in the spin rates of the various galaxies of which it is composed. All this aside, unless the inhabitants of SERPO routinely walk on ceilings and through walls, the laws of physics apply the same there as they do here.

—Lieutenant Colonel ____ ____
USAF Scientific Advisory Group
Comment, November 11, 2005

I've already had independent confirmation from three sources now that the late Dr. Carl Sagan WAS involved in such a project and Project Serpo served as the inspiration for his '85 novel and '97 movie, *Contact*."

—Comment, November 14, 2005

Incidentally, the "senior theory" mentioned above, re: Newton's gravitational attraction law, is only part of the "equation." The other parts are:

1) the equivalence of inertial and gravitational mass, and

(2) the centrifugal force due to orbiting keeps the suns from crashing together and is balanced: M1 omega^2 R1 = M2 omega^2 R2, where omega is the angular rate of rotation (v1/R1 = v2/R2). Each M is the inertial mass that has been shown to equal the gravitational mass and is a basis for Einstein's theory of gravity.

This centrifugal force (each sun pulling on the other) is what results in the rotation about the center of mass at some location along the line joining the two suns. The center of mass is found by using R = R1 + R2 and M1R1 = M2R2, where R is the distance between the suns and the gravitational force between the suns is GM1M2/R^2. When M1 = M2, the center of mass is at the center point of the line joining the centers of the suns. When one mass is huge compared to the other (as with a planet orbiting a sun), the center of mass becomes very close to the center of the most massive sun and the sun "stands still" as the planet orbits. Note that the recent discoveries of planets have been based on the fact that if a planet is heavy enough, the center of mass of the sun-planet system will not be at the center of the sun. In this case the sun orbits the center of mass and this orbiting can produce a doppler shift in the spectrum of light from the sun as the sun goes around the center of mass.

—Comment, November 28, 2005

APPENDIX 5

THE EBEN ENERGY DEVICE

The Crystal Rectangle

The Crystal Rectangle was used by the team as the power source for all their electrical equipment while on Serpo. It was also found at one of the Roswell crash sites and was used to power the apparatus used to communicate with Serpo at Los Alamos. Anonymous replied to a question about the device in "Posting Eleven" on December 21, 2005.

QUESTION: Please explain, for the enlightenment of all, the Eben Energy Device [ED] found at the Roswell crash site.

ANONYMOUS: OK, I will answer this question. Dimensions: 9" x 11" x 1.5", weight 26.7 oz. The ED [Energy Device] is clear and made of something similar to hard plastic. On the bottom left, there is a small square metal plate, possibly a chip. It is one of the connector points. On the bottom right, there is another small square metal point, which is the second connector point. Viewed from an electron microscope, the ED contains small circular-shaped bubbles. Within these bubbles are extremely minute particles. When a demand for electric power is applied to the ED, the particles always move clockwise at a great speed, not measurable.

There is also some type of unidentified fluid located around the bubbles. When a demand is placed on the ED, this fluid turns from a clear color to a hazy pink color. The fluid becomes warm between 102°–115°F. However, the little bubbles would not heat up, ONLY the fluid. The bubbles maintained a constant temperature of 72°F. The boundary of the ED contains small (micron-sized) wires. When a demand is placed on the ED, the wires expand in size. This expanding process depends on the amount of demand placed on the ED. We did extensive, exhaustive experimentation with the ED. We could power everything from a 0.5-watt bulb to an entire house. The ED automatically detects the required demand and then outputs that exact amount. It worked on everything electrical except equipment that contained a magnetic field. Somehow, our magnetic field interferes with the output demand of the ED. However, we have developed a shielding process to correct this.

Additional information about the Crystal Rectangle was provided by a document sent to the Serpo website titled "Classified National Security Document About Pentagen/Crystal Rectangle/Nevada Test Site." It was sent as Release 19 showing no author, and no date. The following was headed "2. UPDATE ON THE CR (Crystal Rectangle—Energy Source)."

A. Since 1956, many experiments were conducted using the CR. Most of the experiments were conducted by Los Alamos or a contractor for the Department of Energy. Remember that the CR was described as follows: The dimensions were 26 cm x 17 cm x 2.5 cm. The CR weighs 728 grams. There is a possibility that "two" CRs exist—one that weighed 668 grams and one that weighed 728 grams. There was a notation in a classified document that read: PVEED-1 [Particle Vacuum Enhanced Energy Device].This would indicate that there is a PVEED-2. Scientists do not refer to the CR as a CR but as a PVEED, or "The Magic Cube."

B. Remember the small dot that moved around the inside when an energy demand was placed on the CR? Our scientists have discovered the substance contained in the dot. The dot was found to be a perfectly rounded particle of charged (?) antimatter. Our scientists still don't understand how this piece of antimatter can remain stable until it is "tasked" with movement. They still don't understand [how] once a demand is made to the CR the antimatter starts its movement and creates energy.

C. Our scientists have found that the CR is made of an unknown material with several unknown elements which have been detected. One of the materials is similar to carbon, but not exactly like carbon as we know it. Another substance is similar to zinc, but not the same consistency of zinc.

D. Our scientists cannot explain the action of the antimatter and the actions of neutrons that are created and then disappear when the demand is lifted.

E. Our scientists cannot explain why the constant temperature of the CR is 72°F. Even when heat is directed on the CR, the temperature remains at 72°F. How this occurs cannot be explained.

F. Some scientists think the CR is operated remotely, perhaps by an unknown satellite in Earth's orbit. However, even [when] shielded, the CR operates normally.

G. When an energy demand is placed on the CR, it creates a signal that can be measured at 23.450 MHZ. However, when increased demand is placed on the CR, the frequency is modulated from 23.450 MHZ to 46.900 MHZ or double the original frequency. However, when the demand is reduced, the frequency drops to 1.25 KHZ, which is a constant frequency when no demand is placed on the CR. Regardless of what power demand is placed on the CR, the frequency NEVER rises to more than 46.900 MHZ!

H. Remember the small set of squares that contained horizontal wires? The wires were determined to be similar to tungsten. The wires somehow conduct energy by bouncing the neutrons off these wires back into the fluid. The small dot bounces against the wire when energy demand is placed on the CR. Remember, only certain wires would react or expand when energy demand was placed on the CR. Scientists think that, depending on the demand, only certain wires would expand. Somehow, the energy output would be controlled by the number of squares used.

I. The U.S. Government [made a] duplicate of the CR. The USG made one in 2001 that actually worked . . . for a short period of time. The operation was EXTREMELY CLASSIFIED and the device blew up at the Nevada Test Site, injuring two (2) employees.

J. The time line for the CR is as follows:
   1) 1947: CR was found in the second crash site.
   2) 1949: Los Alamos scientists first conducted experiments with CR. At this time, no one knew what it was. Some scientists thought it was just a window.
   3) 1954: Sandia Labs conducted several experiments with the CR, but still didn't know its actual use.
   4) 1955: CR was lent to Westinghouse for experiments.
   5) 1958: CR was lent to Corning Glass in an effort to determine its construction material.
   6) 1962: CR's first "official" test [was] conducted at Los Alamos and published in a classified report.
   7) 1970: CR was determined to be more than a window. The CR was found to fit into a space on the spacecraft. Scientists determined [that the] CR was some sort of energy device.
   8) 1978: CR determined to be a high powered energy device that provided electrical power to the spacecraft.
   9) 1982: CR was first tested and produced energy.
   10) 1987: CR was given to E-Systems for extensive testing.
   11) 1990: CR was proven to be an UNLIMITED POWER SYSTEM. The construction and contents of the CR was learned. However, no one knew just how it worked.
   12) 1998: CR project, "The Magic Cube," was started in an effort to accelerate the knowledge of the device.
   13) 2001: CR project, "The Magic Cube" was transferred from Los Alamos' "Futures Division" to its "Special Projects Section K" division.

K. Currently [as of September 2002], the CR is contained at the Section K division facilities, Los Alamos.

APPENDIX 6

THE EBEN PROPULSION SYSTEM

In Release 32, Anonymous answered some miscellaneous questions sent in to the Serpo website. This one concerned Eben space travel technology.

> **QUESTION:** Do we possess teletransporter or "flash drive" technology that we have appropriated/back-engineered from the visitors, yes or no?
>
> **ANONYMOUS:** With respect to your question regarding "teletransporter" technology, the Ebens do NOT have this technology. They have mastered space travel and can venture through space, defying the time barrier. As for our own technology, I very much doubt we have it today. We don't have the *Star Trek* "Beam-me-up-Scotty" technology.

Anonymous turned this question over to a physicist at the DIA, who gave an extensive reply. Bracketed information added by Victor Martinez.

> **QUESTION:** In a private communication I had with your legal counsel, I am given to understand that a member of the DIA-6 has a PhD in [theoretical] physics. . . . Can you ask this individual to expand on HOW SPECIFICALLY the Ebens traverse the vast, great distances of space and

overcome the problem of TIME as they seem to freely travel from Point A to Point B in our Milky Way Galaxy?

ANONYMOUS: The Ebens use a "Universal Grid" system in traveling from one point of space to another. Their craft are able to travel NEAR the speed of light. This enables their craft to go into an altered space-time chamber. That allows the point of DEPARTURE and the point of DESTINATION to become closer in real time. It is similar to folding space by making the two (2) points—departure and destination—become much closer. Remember, the Ebens have been working to perfect this type of space travel—overcoming the time barrier—for well over fifty thousand [of our] years. By now, they have, in fact, perfected this mode of space travel.

Although we have been given the basic BLUEPRINT for their craft, propulsion mechanism, and overall operating system, we still don't understand it. They utilize minerals that we simply don't have here on Earth [Element 115, per Robert Lazar?]. One particular mineral, similar to uranium[3]—but not as radioactive—provides the extra power for their propulsion system. They also utilize a form of SPACE DISPLACEMENT system, which basically causes a VACUUM in front of the propulsion that allows NOTHING to interfere with the created THRUST. At the present time, we canNOT understand how they accomplish this. They use a vacuum chamber, which consists of a mini–nuclear reactor that forces some type of matter into space that deletes the molecules and causes that very small portion of space to become a vacuum. They also utilize antimatter in such a way as to force their propulsion system into "streams" [waves] of energy in front of their craft that enables the craft to move and flow much more easily through space withOUT any FRICTION from the ATMOSPHERE.*

That is all from our physicist member.

Comment Dated November 20, 2005:

In Dan Sherman's book *Above Black,* in which he reveals how he was groomed and trained as a telepathic communicator with a race of ETs whom he received information from and about but never actually

*Presumably, this would not apply to space travel, since there is no atmosphere in space.

met, Sherman states that he learned that "time"—as we humans know it—does not have the same meaning for "them." They still age as we do but are not bound by the physics of time as we currently are. Their means of travel across vast distances is heavily dependent on the manipulation of time, but not as we perceive it.

Sherman asked if they can travel through time: e.g., can they go backward or forward in time? He was told that it was not possible to witness a reality that occurred in some other time but the present. In order to go back in time, one must assume that there exists a reference point from which to measure backward or forward. This is an impossibility. Essentially, they weren't able to travel through time but around time and from time. Sherman says he never really understood what this meant. Their mode of propulsion was somehow that they used "time" and electromagnetic energy. Also they said that our sun was very unique; and that someday we would understand how it really worked and how we could utilize the same methods that they themselves use but on a smaller scale.

Extract from a comment Dated November 22, 2005:

You may know that there is a huge amount of information in the available literature, reportedly coming from contacts with offworld beings, that supports what Dan Sherman states he was told. Specifically, that the nature of "time" is not at all as we think about it. In many cases, the beings have even said that "time" does not exist. I feel they mean for themselves and their technology it does not exist. After all, if you can travel into the past and the future at will, does "time" still exist? If you had a teleporter into which you could step and it would instantly send you to any place on Earth, you might say that (with that technology) "distance" no longer exists. If your transportation time is nearly instantaneous, does distance matter any longer? Conceptually, no! Distance is no longer a barrier to travel and usually our current distance barriers also include the time component for making one's way across that distance. Some feel that time itself is the main factor. If not, why would we be building faster and faster jets and planning hypersonic craft to get us there even more quickly?

One case I've studied extensively says the beings are capable of near-instantaneous jumps in distance that we simply cannot imagine. They claim that once technology reaches the point where one can access a certain "extra-dimensional" area of the universe, travel over millions of light-years is accomplished in the blink of an eye. Our science fiction can imagine that, but most of our real minds simply cannot believe it. But, if true, as in the teleporter example above, then distance and time become just concepts and no longer barriers. So distance and time would "no longer exist." Obviously, that's a technicality, a concept, because the miles between points are still the same—but the "time" it takes to cross the distance has essentially dissolved. Some type of system for measurement for crew changes, sleep periods, meeting times, etc. would still need to exist, simply for convenience and scheduling sake, so things operate in some sort of sync.

Excerpt from Release 26A taken from *The Day after Roswell,* by Col. Philip J. Corso:

There were no conventional technological explanations for the way the Roswell craft's propulsion system operated. There were no atomic engines, no rockets, no jets, nor any propeller-driven form of thrust. The craft was able to displace gravity through the propagation of magnetic wave[s], controlled by shifting the magnetic poles around the craft so as to control, or vector, not a propulsion system, but the REPULSION FORCE of like charges.

Once they realized this, engineers at our country's primary defense contractors raced among themselves to figure out how the craft could retain its electric capacity and how the pilots who navigated it could live within the energy field of a wave. The initial revelations into the nature of the spacecraft and its pilot interface came very quickly during the first few years of testing at Norton AFB. The Air Force discovered that the entire vehicle functioned just like a giant capacitor. In other words, the craft itself stored the energy necessary to propagate the magnetic wave that elevated it, allowed it to achieve escape velocity from the Earth's gravity, and enabled it to achieve speeds of over 7,000 mph.

The pilots weren't affected by the tremendous g-forces that build up in the acceleration of conventional aircraft because to aliens inside, it was as if gravity was being folded around the outside of the wave that enveloped the craft. Maybe it was like traveling inside the eye of a hurricane. But how did the pilots interface with the wave form they were generating?

The secret to this system could be found in the single-piece skin-tight coveralls spun around the creatures. The lengthwise atomic alignment of the strange fabric was a clue that somehow the pilots became part of the electrical storage and generation of the craft itself.

They didn't just pilot or navigate the vehicle . . . they became part of the electrical circuitry of the vehicle, vectoring it in a way similar to the way you order a voluntary muscle to move. The vehicle was simply an extension of their own bodies because it was tied into their neurological systems in ways that even today we are just beginning to utilize.

So the creatures were able to survive extended periods living inside a high-energy wave by becoming the primary circuit in the control of the wave. They were protected by their suits, which enclosed them head to feet, but THEIR SUITS ENABLED THEM TO BECOME ONE WITH THE VEHICLE, literally part of the wave.

In 1947, this was a technology so new to us that it was as frightening as it was frustrating. If we could only develop the power source necessary to generate a consistently well-defined magnetic wave around a vehicle, we could harness a technology by which we would have surpassed all forms of rocket and jet propulsion. It's a process we're still trying to master today, fifty years [sixty years now] after the craft fell into our possession.*

*As pointed out in chapter 3, Colonel Corso was not privy to the reverse-engineering of alien craft that had been going on since 1953. He did not have a high enough clearance. When he wrote his book in 1997, we already had working antigravity craft.

APPENDIX 7

EBEN RELIGIOUS BELIEFS

Below is an excerpt from Release 28, undated, wherein Anonymous discusses a meeting between an Eben visitor he refers to as OSG (Our Special Guest) and Pope Benedict XVI in Washington, D.C. The meeting took place during the Pope's visit in April, 2008, followed by meetings with Vatican representatives at Groom Lake. These visits were opportunities to compare Eben spiritual views and beliefs with Catholicism.

> During each visit, a representative from the Vatican WAS present. The pope was particularly interested in the religious activity of the Ebens. The Ebens worship a God. The pope feels their God is the same as ours. The Ebens worship God differently, but NOT so much. In fact, OSG ["Our Special Guest" referring to an Eben "ambassador"] brought artifacts of the Ebens' God that fits directly into OUR [Christian] God.
>
> Several Eben paintings, sculptured statutes, and carved fetishes were similar to our God. In fact, the story of their God—appearing thousands of years ago on SERPO and setting up religious sects on their planet—is similar to our story of Jesus. The Ebens chant verses, which, when translated, are similar to OUR prayers. The Eben chants contain twenty-six verses, which they repeat every day at their prayer hour, which is in the afternoon (SERPO day). The chants sound like Tibetan chants. On a particular Eben day of their year, the Ebens expand the chants to

thirty-eight verses. The extra twelve chants pertain to "angels"—which we have translated to mean "saints"—who have helped the Eben society. This information has NEVER been released.

The basic beliefs/religion of the Ebens are simple. However, their practice is very complex. The Ebens worship one God, which they call an "Entity," and they have religious symbols that reflect other religious entities, which they call "subentities." This would be similar to our identification with saints.

The Ebens' belief in life after death is similar to that of the Roman Catholic church and some Eastern religious doctrines. Once an Eben dies, [his or her] soul [bioplasmic body] is taken from the body by these subentities (saints) and cleansed of all sins. The soul is then taken to a midpoint (Catholics would call it "Purgatory") between Heaven and that midpoint. Then once the soul is ready, it is taken to the "Supreme Plateau" (Heaven), where it remains for an eternity.

Their beliefs become much more complex at this point. Some souls, called "The Arranged" (that is their word), are prepared for entry back into the living society, i.e., this plane of existence. The Ebens believe that if they perform some specific act—referred to as "karma" on Earth—during their regular life, they then can come back to the living life in another body. The Ebens believe in reincarnation and the eternity of the soul. The Ebens do NOT believe that animals or their sworn enemies of other space-traveling races have souls.

This is something that might help hold people over on the "Project SERPO" subject matter until all of our files and materials have been given back to us from the Dept of Defense.

APPENDIX 8

THE DEFENSE INTELLIGENCE AGENCY

By Victor Martinez

This paper by Victor Martinez, the Serpo website moderator, provides some in-depth information about the DIA that is necessary to understand their dedication to transparency, which is unique among government intelligence agencies.

The DEFENSE INTELLIGENCE AGENCY is organized into six (6) directorates and the Joint Military Intelligence College (formerly the Defense Intelligence College). The directorates are:

Administration
Analysis
Human Intelligence [HUMINT]
Information Management and Chief Information Center
Intelligence Joint Staff
Measurement and Signature Intelligence [MASINT] and Technical Collection

The DIA headquarters is in the Pentagon. The DEFENSE INTELLIGENCE ANALYSIS CENTER is an extension of that headquarters which is at Bolling Air Force Base in southwest Washington, D.C., as is the Joint Military Intelligence College. [Bolling AFB is where all of the Project

Serpo files are located, which include thousands of photographs of the Eben civilization in several large photo album books; animal, plant and soil samples; audio recordings of the Eben music; and photos of other alien species that visited/were cloned on Serpo.] A few DIA employees are based at the Armed Forces Medical Intelligence Center in Maryland, and at the Missile and Space Intelligence Center in Alabama. The DIA's military attachés are also assigned to U.S. embassies around the world and as liaison officers to each unified military command. The DIA's Russian counterpart, or parallel, organization is the *Glavnoye Razvedyvatelnoye Upravleniye,* or GRU [Main (Chief) Intelligence Administration].

PRELUDE: HOW AND WHY IT ALL STARTED

The creation of a unified Dept of Defense [DOD] in 1947–49 was not accompanied by the unification of defense activities. Each of the military services maintained its one [own] intelligence organization; indeed, maintaining these distinct capabilities had been a major demand of the military during deliberations over the creation of the CIA. But there were also a number of intelligence requirements that were either interservice or departmentwide. Thus, an additional intelligence organization had to be designed and developed to meet these broader, growing needs for the future.

THE BIRTH OF THE DIA

The U.S. Dept of Defense established the DIA on Sunday, October 1, 1961, to coordinate the intelligence activities of the military services. The DIA serves as the intelligence agency for the Joint Chiefs of Staff [JCS] as well as for the Secretary of Defense and the U.S. unified or theater military commanders. As a senior military intelligence component of the U.S. intelligence community, the DEFENSE INTELLIGENCE AGENCY is a combat-support arm, providing all-source intelligence to the American armed forces, defense policy makers, and other members of the U.S. intelligence community. Under the Defense Reorganization

Act of 1958, several unified military commands were created, but as long as each individual service had its own intelligence organization, the unified commands would not be receiving unified intelligence. The services had set up barriers that prevented the free exchange of intelligence data.

Complaints about vastly different estimates and bureaucratic infighting inspired the creation of the DIA under the KENNEDY administration, although the specific efforts by the Dept of Defense "to put its house in order" with respect to intelligence dates back to 1959. In his first State of the Union message, President Kennedy said: "The capacity to act decisively at the exact time action is needed has too often been muffled, [creating] a growing gap between decision and execution, between planning and reality."

The brainchild of President JOHN F. KENNEDY and his secretary of defense, ROBERT S. McNAMARA, the DIA was established in 1961 as a military intelligence authority that would provide independent information while circumventing the "turf" problems arising from interservice rivalries. [See end of this section.]

The Secretary of Defense ROBERT S. McNAMARA created the DIA, giving to it as a prime mission the coordinating of intelligence estimates, which previously had been individually produced by the individual services. The DIA is a member of the intelligence community and, as such, in theory comes under the nominal responsibilities of the director of central intelligence [DCI] as well as the secretary of defense. Further, as originally set up, the DIA director assumed the functions of the J-2 (intelligence) within the JCS; the DIA still provides support for the J-2. The classified Plan for the Activation of the Defense Intelligence Agency (1961) called for a maximum of 250 personnel—military and civilian—at DIA headquarters.

THE DEPARTMENT OF DEFENSE IS REVAMPED

The military services retained their intelligence agencies—Air Force intelligence, Army intelligence, Naval intelligence—and responsibilities for intelligence training, developing doctrine for combat intelligence, internal security, and counterintelligence within their respective services. Other duties retained by the services, but available to the DIA for its mission, included the collection of technical intelligence and intelligence support for JCS studies.

The DIA has frequently struggled—through reorganizations and Pentagon lobbying—to increase its importance within the intelligence community. But in fulfilling its charter to collect military and military-related intelligence, the DIA must rely upon the National Reconnaissance Office [NRO] for military information obtained by satellites and strategic reconaissance [*sic*] aircraft; the National Security Agency [NSA] for the making and breaking of codes; and the CIA for military intelligence gained from foreign intelligence agencies. If, for example, the CIA "turns" a Russian GRU officer, the DIA must depend on the CIA to obtain the GRU officer's information.

By 1975, the DIA had more than 4,600 employees and an annual budget estimated at more than $200 million. However, the DIA had been among the intelligence agencies most severely hit by the end of the Cold War, which led to a 25 percent reduction in its personnel. Former DCI Admiral STANSFIELD TURNER wrote in 1986: "Because the DIA is self-conscious about living within the shadow of the more capable CIA, it often takes contrary positions just to assert its independence. . . . More often than not, when the DIA does produce a differing view, it cannot—or will not—support it." Turner, like many other CIA officials down the years, also criticized the DIA for being unable to dominate with the competing military services.

There were great changes in intelligence network and routing systems to battlefield and at-sea commanders. In February 1991, the DIA began producing a closed-circuit telecast to about 1,000 defense intelligence and operations officers in the Pentagon and at 19 military commands in the United States.

THE DIA IN THE MODERN ERA

The *Defense Intelligence Network Show* is encrypted so that it can be watched ONLY by authorized monitors. Ingredients for the telecasts include aerial and satellite reconnaissance images and audio reports from the NSA. "We've got to do to intelligence what CNN has done to news," a Pentagon official told The WASHINGTON POST.

The DIA has also provided intelligence to United Nations peacekeeping forces and to U.S. responses to terrorist actions. The DIA also aids law enforcement agencies involved in antidrug operations. There has been a marked improvement in performance as the military establishment has changed its attitude toward intelligence, which had been seen as a dead end for nonspecialist officer career paths.

Although the DIA was conceived as a military agency, by the mid-1980s, about 60% of the DIA staff were civilians. The DIA has sometimes found itself torn between its military customers (the JCS and their organization) and the civilian customers of the DOD. The Joint Chiefs may seek analysis to support specific or preferred positions; the civilians may prove skeptical of military-produced analysis, which often tends toward more pessimistic assumptions about conflict and combat.

A possible renaissance [*sic*] for the DIA came in 1995 with the appointment of JOHN M. DEUTCH, former deputy secretary of defense, as DCI. During his time in the Pentagon, Deutch had taken a close interest in the DIA and had created within it the Defense HUMINT Service [DHS], which is authorized to run agents and proprietary companies overseas.

After the capture of Saddam Hussein in December 2003, the CIA was chosen to lead his interrogation. But specialists in the DIA, who had operated extensively in Iraq, were also part of the interrogation team. DIA analysts were also involved in the hunt for weapons of mass destruction.

DIA VS CIA: "SIBLING RIVALRY" AND INTELLIGENCE TURF WARS

The so-called "sibling rivalry" is just that: it is a CIA term used to refer to officers employed by the sometimes rival DIA. The UNofficial rivalry between the two (2) agencies began when the DIA was established in 1961. From the beginning, some CIA officials felt that the DIA was encroaching on agency territory. It was believed that the DIA was too involved with CIA-controlled spy satellite operations. The rivalry also stemmed from fiscal concerns, wherein both agencies found themselves competing for budget dollars. However, by virtue of the coordinating and oversight authority of the DCI, the CIA is senior to the DIA within the U.S. intelligence community. Today, the DIA very effectively reduces the role of the individual armed services in the realm of strategic intelligence.

APPENDIX 9

DEEP SPACE PROBES

In addition to the publicized space probes such as *Voyager,* the U.S. has been sending other probes into deep space since 1965 that have been kept secret. We now learn from Anonymous that the main purpose of these probes is to set up a reliable communication system with Serpo. Clark McClelland, former NASA Aerospace Engineer, said in 1999: "Some of these NSA Probes are launched from the Kennedy Space Center (KSC) during secret missions called 'Classified' missions. They loaded the payloads on the Space Shuttle at night under tight security and few technicians had the proper clearances to participate. An all-male crew of specially trained military astronauts flew those missions."

Anonymous gives us the following information about these probes:

> NSA/NASA both teamed up to develop new technologies to explore the universe. They have deployed the following deep space probes. These probes were used to establish a communication link with the ALIENS. They formed a type of repeater system for the communications. Not much else is known.

The following is a list of known probes:

A. 1965: First deep space probe, Code name: "Patty"
B. 1967: Second deep space probe, Code name: "Sween"
C. 1972: Third deep space probe, Code name: "Dakota"
D. 1978: Fourth deep space probe, Code name: unknown

E. 1982: Fifth deep space probe, Code name: unknown
F. 1983: Sixth deep space probe, Code name: unknown
G. 1983: Seventh deep space probe, Code name: unknown
H. 1983: Eighth deep space probe, Code name: "Moe"
I. 1985: Space probe launched on SS Mission 51-J, Code name: "Sting Ray"
J. 1988: Ninth deep space probe, Code name: "Amber Light"
K. 1988: Tenth deep space probe, Code name: "Sandal Slipper"
L. 1989: Eleventh deep space probe, Code name: "Cocker Peak"
M. 1992: Twelfth deep space probe, Code name: "Twinkle Eyes"
N. 1997: Thirteenth deep space probe, Code name: "Kite Tangle"

APPENDIX 10

BACK-ENGINEERED ALIEN CRAFT

A Disclosure Project Statement

Under the aegis of "The Disclosure Project," Dr. Steven Greer interviewed and recorded hundreds of insider witnesses to, and participants in, top-secret UFO and extraterrestrial procedures and events. This video interview with the late Captain Bill Uhouse is on YouTube (www.youtube.com/watch?v=VzGYNJEA4Ag&feature=youtu.be), where Uhouse is labeled "Witness #2." It was conducted in October, 2000. This is a critical film because it validates the fact that alien scientists helped us to develop antigravity craft at Area 51 beginning in 1953, using the Kingman craft as a model, as discussed in chapter 5.

Bill Uhouse says:

> I spent ten years in the Marine Corps, and four years working with the Air Force as a civilian doing experimental testing on aircraft since my Marine Corps days. I was a pilot in the service, and a fighter pilot; [I] fought in the latter part of World War II and the Korean War. I was discharged as a captain in the Marine Corps.
>
> I didn't start working on flight simulators until about—well, the year was 1954, in September. After I got out of the Marine Corps, I took a job with the Air Force at Wright-Patterson doing experimental flight-testing on various different modifications of aircraft.

While I was at Wright-Patterson, I was approached by an individual who—and I'm not going to mention his name—[wanted] to determine if I wanted to work in an area on new creative devices. Okay? And, that was a flying disc simulator. What they had done: they had selected several of us, and they reassigned me to Link Aviation, which was a simulator manufacturer. At that time they were building what they called the C-11B, and F-102 simulator, B-47 simulator, and so forth. They wanted us to get experienced before we actually started work on the flying disc simulator, which I spent thirty-some years working on.

I don't think any flying disc simulators went into operation until the early 1960s—around 1962 or 1963. The reason why I am saying this is because the simulator wasn't actually functional until around 1958. The simulator that they used was for the extraterrestrial craft they had, which is a thirty-meter one that crashed in Kingman, Arizona, back in 1953 or 1952. That's the first one that they took out to the test flight. This ET craft was a controlled craft that the aliens wanted to present to our government—the USA. It landed about fifteen miles from what used to be an Army Air base, which is now a defunct Army base. But that particular craft, there were some problems with: number one—getting it on the flatbed to take it up to Area 51. They couldn't get it across the dam because of the road. It had to be barged across the Colorado River at the time, and then taken up Route 93 out to Area 51, which was just being constructed at the time. There were four aliens aboard that thing, and those aliens went to Los Alamos for testing.

They set up Los Alamos with a particular area for those guys, and they put certain people in there with them—people who were astrophysicists and general scientists—to ask them questions. The way the story was told to me was: there was only one alien that would talk to any of these scientists that they put in the lab with them. The rest wouldn't talk to anybody, or even have a conversation with them. You know, first they thought it was all ESP or telepathy, but you know, most of that is kind of a joke to me, because they actually speak—maybe not like we do—but they actually speak and converse. But there was only one who would [at Los Alamos].

The difference between this disc and other discs that they had looked at was that this one was a much simpler design. The disc simulator didn't have a reactor, [but] we had a space in it that looked like the reactor that wasn't the device we operated the simulator with. We operated it with six large capacitors that were charged with a million volts each, so there were six million volts in those capacitors. They were the largest capacitors ever built. These particular capacitors, they'd last for thirty minutes, so you could get in there and actually work the controls and do what you had to—to get the simulator, the disc, to operate.

So, it wasn't that simple, because we only had 30 minutes. Okay? But in the simulator you'll notice that there are no seat belts. Right? It was the same thing with the actual craft—no seat belts. You don't need seat belts, because when you fly one of these things upside down, there is no upside down like in a regular aircraft—you just don't feel it. There's a simple explanation for that: you have your own gravitational field right inside the craft, so if you are flying upside down—to you—you are right side up. I mean, it's just really simple, if people would look at it. I was inside the actual alien craft for a start-up. There weren't any windows. The only way we had any visibility at all was done with cameras or video-type devices. My specialty was the flight deck and the instruments on the flight deck. I knew about the gravitational field and what it took to get people trained.

Because the disc has its own gravitational field, you would be sick or disoriented for about two minutes after getting in, after it was cranked up. It takes a lot of time to become used to it. Because of the area and the smallness of it, just to raise your hand becomes complicated. You have to be trained—trained with your mind, to accept what you are going to actually feel and experience.

Just moving about is difficult, but after a while you get used to it and you do it—it's simple. You just have to know where everything is, and you [have] to understand what's going to happen to your body. It's no different than accepting the g-forces when you are flying an aircraft or coming out of a dive. It's a whole new ball game.

Each engineer who had anything to do with the design was part of the start-up crew. We would have to verify all the equipment that we put in—be sure it [worked] like it [was] supposed to, etc. I'm sure our crews have taken these craft out into space. I'm saying it probably took a while to train enough of the people, over a sufficient time period. The whole problem with the disc is that it is so exacting in its design and so forth. It can't be used like we use aircraft today, with dropping bombs and having machine guns in the wings.

The design is so exacting, that you can't add anything—it's got to be just right. There's a big problem in the design of where things are put. Say, where the center of the aircraft is, and that type of thing. Even the fact that we raised it three feet so the taller guys could get in—the actual ship was extended back to its original configuration, but it has to be raised.

We had meetings, and I ended up in a meeting with an alien. I called him J-Rod—of course, that's what they called him. I don't know if that was his real name or not, but that's the name the linguist gave him. I did draw a sketch, before I left, of him in a meeting. I provided it to some people and that was my impression of what I saw, an art picture of an alien that is working in cooperation with Earth-people as told here.

The alien used to come in with [Dr. Edward] Teller and some of the other guys, occasionally, to handle questions that maybe we'd have. But you have to understand that everything was specific to the group. If it wasn't specific to the group, you couldn't talk about it. It was on a need-to-know basis. And [the ET], he'd talk. He would talk, but he'd sound just like as if you spoke—he'd sound like you. You know, he's like a parrot, but he'd try and answer your question. A lot of times he'd have a hard

Sketch of J-Rod wearing a human man's shirt. Drawing by retired mechanical engineer Bill Uhouse, based on the entity's appearance at a science meeting with physicist Edward Teller and other scientists in the 1970s or early 1980s.

time understanding, because if you didn't put it on paper and explain yourself, half the time he couldn't give you a good answer.

The preparation we had before meeting this alien was, basically, going through all of the different nationalities in the world. Then they got into going into other forms of life, even down to animals and that type of thing. And this J-Rod—his skin was pinkish, but a little bit rough—that kind of stuff; not horrible-looking, you know—or, to me, he wasn't horrible-looking.

Some of the guys who were in the particular group that I was in—they never even made it. . . . [W]hen they gave you the psychological questions, I just answered them the way I felt and I had no problem. That's what they wanted to know—if you'd become upset—but it never bothered me. It didn't amount to much. So basically, the alien was only giving engineering advice and science advice. For example, I performed the calculations but needed more help. I spoke of a book that—well, it's not a book; it's a big assembly with various divisions dealing with gravitational technology, and the key elements are in there, but all the information wasn't there. Even our top mathematicians couldn't figure some of this stuff out, so the alien would assist.

Sometimes you'd get into a spot where you [would] try and try and try, and it wouldn't work. And that's when he'd come in [the alien]. They would tell him to look at this and see what we did wrong. Over the last 40 years or so, not counting the simulators—I'm talking about actual craft—there are probably two or three dozen, and various sizes that we built.

I don't know much about the ones that they brought here [ETs]. I know about that one [craft] out of Kingman, but that's about it. And I know the company that hauled it out of there—who is out here now. But, there's one that operates with certain chemicals.

I think these triangles that people are seeing are two or three t30-meter craft that are in the center of it [the triangle]. And the outside perimeter—well, you could put anything you want, as long as these particular ones meet the design criteria, and they'll operate.

You know, there were certain reasons for the secrecy. I could understand that; it was no different than the first atomic bomb that they built.

But they are getting so far ahead now with aircraft design. And like I told you gentlemen earlier—that by 2003, most of this stuff will be out for everybody to look at. Maybe not the way that everybody expects it, but in some manner they determine appropriate to show everybody. You know, a big surprise. The reason why I said that is because the document I signed ends in 2003 and I'm not the only one who signed those. But that gravitational manual—if you ever get one of these volumes of documents, you'd be on top of the world. You'd know everything.

APPENDIX 11

EXCERPT FROM PRESIDENT REAGAN'S BRIEFING OF MARCH 1981

In this appendix, the CARETAKER resumes his briefing to President Ronald Reagan. In chapter 3, which was about Roswell, he informed the president about the crash and its ramifications. Here, he gives the details of Project Serpo.

The CARETAKER: Mr. President, in 1964, we were able to have our very first controlled encounter with the Ebens. Let me first give you the background. EBE was a mechanic, not a scientist. He was still able to teach us some of the Eben language. Their language was very difficult for our linguists to learn because it consisted of tones, not words. However, we were able to translate some basic words. EBE showed us their communications device. It was a strange-looking device that had three (3) parts. Once assembled, the device sent out signals, something like our Morse code system, although there was a problem. During the crash in 1947, one part of this communication system was broken. EBE was unable to repair it until our scientists found some items that could be used in place of the broken parts. Once the communication device was repaired, EBE sent our messages. We had to trust EBE as to the contents of those messages.

You can imagine what some of our military commanders thought of this. EBE could be sending out a distress call that could result in some invasion. But that, of course, never happened. EBE continued to send messages until his death. But once he died, then we were on our own. We were able to crudely operate the device. We sent several messages out over a six-(6) month period (1953). But we did not receive any return messages.

PRESIDENT: Excuse me, did EBE receive any return messages?

The CARETAKER: Getting back to the messages, Mr. President, EBE sent out six (6) messages. One letting his home planet know that he was alive and his comrades were dead, another explaining the two crashes, the third was a request to be rescued, the fourth was a message suggesting a meeting between his leaders and our leaders. The last message suggested some form of an EXCHANGE program.

WM CASEY: Mr. President, we'll go into that later.

PRESIDENT: (not understood). . . What . . . the exchange program?

WM CASEY: Yes, Mr. President. We can give you another couple of hours on that subject.

PRESIDENT: We had one?

WM CASEY: Can I speak to you privately, please, Mr. President?

PRESIDENT: OK, yes . . . you mean now? (not understood)

WM CASEY: Well, let us put this one on the back burner and go on with the remainder of this briefing.

PRESIDENT: OK.

The CARETAKER: Mr. President, we don't think he did, but we could not be entirely certain. But, our scientists fine-tuned our efforts over the next eighteen months and finally sent two messages in 1955 that were received. We received a reply. We were able to translate about 30 percent of the message. We turned to several language specialists from several different universities and even several from foreign universities. Finally, we were able to translate most of the messages. We decided to reply in English and see if the Ebens could translate our language more easily than we could theirs.

PRESIDENT: What did the messages say? The one we received from the Ebens? So, I guess they didn't get the messages sent by EBE? Or did it take that long to respond? Oh, yes, EBE died before we got those messages, never mind.

The CARETAKER: Mr. President, the first message we received acknowledged our message and asked questions about the crew of the two missing craft. It also gave a series of numbers that we think were some type of coordinates.

PRESIDENT: OK, so they wanted to know the coordinates of the crash sites on Earth? I'm sure they wanted to know about their crew. Did we tell them all but one was dead? No, wait; I'm sure when EBE sent his messages that is probably the first thing he sent. Was EBE a military person or what?

The CARETAKER: Mr. President, we believe EBE was a member of their air force or maybe something like NASA.

PRESIDENT: OK, please continue.

The CARETAKER: Thank you, Mr. President. Finally, we were able to translate most of the messages. As I said, we decided to respond in English. Approximately four months later, we received a reply in broken English. Sentences contained nouns and adjectives, but no verbs. It took us several months to translate the message. We then sent Eben our typed English lessons in a series of one-sheet formats.

Without going into the technical description of the Eben communications device, it was like a television screen and a keypad, but the pad contained several different Eben characters depending on the number of times you held down one key. We were able to transpose our English-typed words into the second part of the device, which was similar to our facsimile transmission system. It took our scientists some time to perfect this, but it worked. Six months later, we received another English message. This time it was clearer, but not clear enough. Ebens were confusing several different English words and still failed to complete a proper sentence.

PRESIDENT: Gee, I do that all of the time (sounds of laughter). I just cannot imagine how an alien race could view our language. We have thousands of different languages on Earth and they probably have just one on their SERPO planet. That is truly amazing.

The CARETAKER: Yes, Mr. President, I cannot imagine living on a planet with just one language. But we were able to provide the basic skill level for them to communicate in English. It took time, but they realized our efforts. In one message, they provided us with a form of the Eben alphabet with the equivalent English letters. Our linguists had a very difficult time figuring this out. The written Eben language was simple characters and symbols, but our linguists had a difficult time comparing the two written languages.

Over the next five (5) years, we were able to perfect our understanding of the Eben language somewhat and the Ebens were able to better understand English. However, we had a major problem—trying to coordinate a date, time, and location for an Eben landing on Earth. Even though we could basically understand some Eben and the Ebens could understand some English, we could not understand their time and date system and they could not understand ours. We sent them our Earth's rotation schedule, revolution, date system, etc.

For some reason the Ebens never understood this. In return, the Ebens sent us their system, which was difficult for our scientists to understand because we had no reference to their planet. The Ebens did not explain any astronomical date of SERPO or their system. We then decided to just send pictures showing Earth, landmarks, and a simple numbering system for time periods. We had many problems trying to send pictures using their facsimile system. We couldn't be sure they were receiving what we sent.

We had a lot of trial and errors in doing this. We received back some strange messages from the Ebens, basically big question marks regarding what we sent them as to the pictures. We then decided on narrowing any future landing location for them to the location of their crash in

New Mexico. We concluded they must have that location. We are sure EBE sent that to his home planet prior to his death. We did find some star charts . . . well . . . as we call them, in both crashed spacecraft.

They were difficult to understand because they were on a block that we later figured out went into a certain panel on the crashed craft's instruments. Once the panel was in place, the board showed a star system. In fact, we were able to fit all the found boards into the panel and view many different star systems. We then put to work our astronomers in deciphering the star systems. It didn't take them very long to determine the various star systems. We also found several strange spots on the star charts.

We concluded these spots were where the travel space tunnels that EBE described were located. Our astronomers compared the different star charts and found that they were not consecutive. Meaning that one star chart was from one part of the universe and the next was a chart closer to their home system. Our scientists concluded the spots on the chart were a form of shortcuts from one point of space to another. Some of our top astronomers were briefed into the program in order to study the charts. I'm sure they were given only the minimum amount of information they needed, something like a need-to-know program.

PRESIDENT: OK, that is a lot to absorb. Wow, well, I have many questions, but I guess I'll just wait now. I have something to attend to now. But let us take a short break and come back to this.

WM CASEY: Mr. President, how much time do you have left?

PRESIDENT: Well, Bill, let me check. (Long pause.) I need to call some people on another matter. Give me about fifteen minutes. Is that OK?

WM CASEY: Yes, Mr. President, we are here at your disposal.

PRESIDENT: I have listened intently to this briefing. I have many questions, which I realize traverses several different layers of secrecy. I don't want to mix up the different layers. But I can see how government bureaucracy exists. That is one thing I can probably change as president! Bill, let's go to the next layer.

WM CASEY: Mr. President, do you want the same people involved?

PRESIDENT: Yes, let's just continue.

WM CASEY: OK, CARETAKER, take over.

The CARETAKER: Thank you. When EBE was alive, he showed us two devices. One was a communication system and one was an energy device. The communication system did not work without the energy device. Eventually, a scientist from Los Alamos figured out the two systems and connected them. After EBE died, we were able to send transmissions, as I said earlier. EBE built up a strong friendship with a U.S. Army major, who was his guardian.

The two of them decided that one of Eben's first messages (of the five sent) was a request for an exchange program between the Ebens and our military personnel. Remember, I mentioned six (6) messages. The sixth consisted of landing coordinates for Earth. That information wasn't clearly documented back then. We are not sure of the exact chain of events between EBE and the major. As I said earlier, we were able to eventually communicate with the Ebens.

Over a period of a few years, we could send and receive information. We finally received a startling message from the Ebens. They wanted to visit Earth, retrieve their spacemen['s] bodies and meet with Earthlings. They provided a time, date, and location. We figure that the Ebens were continually visiting Earth and had probably mapped it. However, the date was about eight (8) years in the future. Our military figured something was wrong and that maybe the Ebens were confusing Earth time with Eben time. After a long series of messages, it was determined the Ebens would land on Earth on Friday, April 24, 1964.

PRESIDENT: Just how did we figure the date?

The CARETAKER: Mr. President, these messages occurred over a period of several years. By this time, we both had a working knowledge of the other's seasons, which was based on the Earth's rotation, which also figured into our time periods. We had a working knowledge of their forty-hour days. They were a little smarter than us, being able to comprehend our language and our time periods.

PRESIDENT: OK, that makes sense. But . . .(not understood). . . about . . . (not understood). . . the aliens?

The CARETAKER: Mr. President, we did have a basic understanding of their language. We could understand basic words and symbols. They understood more of our language than we did theirs.

PRESIDENT: OK, then what happened?

The CARETAKER: Well . . .

WM CASEY: Mr. President, this is where things get very interesting.

PRESIDENT: OK, I'm waiting . . . (not understood)

The CARETAKER: Our government, specifically MJ-12, met in secret to plan the event. Decisions were made, then changed many times. We had just about twenty-five months from the time we finally received their message of the date to prepare for their arrival. Several months into the planning, President Kennedy decided to approve a plan to exchange a special military team. The USAF was tasked as the lead agency.

The USAF officials picked special civilian scientists to assist in the planning and crew selection. The team members' selection process was the hardest to accomplish. Several plans were suggested and then changed. It took months for the planners to decide on the selection criteria for each team member. They decided that each member must be military, single, no children, and a career member. They had to be trained in different skills.

WM CASEY: CARETAKER, let's just go into the general stuff here. I don't think the president wants to know every single minute detail.

PRESIDENT: Well, if I had the time, I would (not understood) . . . but, I understand that.

The CARETAKER: Mr. President, a team of 12 men were selected. However, during this time period, President Kennedy died. The nation was shocked, as you know . . .

PRESIDENT: Yes, everyone was shocked. I can understand what must have happened during the project when John died.

The CARETAKER: President Johnson continued the program. When it came time for the meeting, we were ready. The landing occurred in New Mexico. We had everything prepared. We had a hoax landing location just in case it was leaked. The landing occurred and we greeted the Ebens. However, a mixup happened. They were not prepared to accept our exchange personnel. Everything was placed on hold. Finally, in 1965, the Ebens landed in Nevada and we exchanged 12 of our men for one of theirs.

PRESIDENT: One? Why just one?

WM CASEY: Mr. President, this wasn't clearly documented in the reports that we read.

PRESIDENT: One . . . was this their ambassador?

WM CASEY: Well, something like that. We just called it Ebe2. We'll discuss that later.

The CARETAKER: Mr. President, our team of twelve went to the Eben planet for thirteen years. The original mission called for a ten-year stay, however, because of the strange time periods on their planet, the team stayed three additional years. Eight [seven] returned in 1978. Two died on the planet and two decided to stay.

[***Note:*** Team Member #308 (team pilot #2) died of a pulmonary embolism en route to SERPO on the nine-month journey; eleven arrived safely.]

PRESIDENT: OK, this is just AMAZING! I can see, about that movie. The movie was based on a real event. I saw that movie. Twelve men left, along with Richard Dreyfuss. [*Close Encounters of the Third Kind,* 1977]

WM CASEY: Mr. President, yes, the movie was similar to the real event, at least the last part of the movie.

APPENDIX 12

A FRAMEWORK FOR PUBLIC ACCLIMATION

The following document was presented to Victor Martinez from another source (not Anonymous) and was posted to the Serpo website. If it did indeed originate from MJ-12, then it comes from the very pinnacle of the classified hierarchy. It is a remarkable document. In defining the goals of the Public Acclimation Program, it reveals in twelve short declarations everything we have secretly learned about extraterrestrials.

MJ-12 Staff Document, A FRAMEWORK: 30 July, 1999

FACTS CONCERNING EXOBIOLOGICAL LIFE TO BE CONVEYED THROUGH "THE PUBLIC ACCLIMATION PROGRAM"

1) Intelligent life does exist on other planets and throughout the universe.
2) Craft not of human design or manufacture are operating in and around the land, sea, and air of Earth.
3) Intelligent beings other than Homo Sapiens are conducting various missions on this planet. These beings have been coming here for tens of thousands of years.
4) Alien beings may have humanlike bodies or nonhuman bodies (such as hybrid, insectoid, or reptilian). Intelligent beings can be physical, nonphysical, or interdimensional in nature.

5) The variety of life in the universe is diverse, like the life on our own planet is diverse.
6) Some alien beings have the ability through advanced technology or other means to move forward and backward at will through time and space.
7) The spiritual evolution of an alien life-form may be ahead of, equal to or behind its level of technological development.
8) The social orientation, motives, and agendas of these beings is very diverse. Some alien intelligences are more friendly to human beings than others.
9) In many cases, the "abduction phenomenon" is a real event. This activity is complex, coordinated and purposeful. It often occurs throughout many generations of a family.
10) Crossbreeding of humans with more than one alien species has occurred. Hybrid children and hybrid adults do exist. They have characteristics of both the human and alien races.
11) Although most alien contacts and sightings occurring on modern-day Earth have been shrouded in secrecy and mystery, the veil is slowly being lifted by the activities of civilians and specially assigned government personnel. The Public Acclimation to the reality of alien life is proceeding in a way designed not to shock or disrupt society any more than necessary.
12) A great amount of "UFO" and alien information is now in the public domain. Countless books, videos, and Internet websites are devoted to these subjects. Thousands of pages of U.S. government documents on unusual sightings and encounters have been made available.

At this point, credentialed researchers and academicians have investigated the UFO/alien phenomenon and published their findings. This is a further step to help validate the phenomenon in the minds of the public. With these twelve points as a reliable framework, it is intended that key members of the public and government will be better able to accept, evaluate, and place into perspective the large body of evidence which is soon to be before them.